INTERMEDIATE

Practice Book

KU-226-905

MASTERING MATHEMATICS

FOR **WJEC** GCSE

Practice ● Reinforcement ● Progress

Assessment Consultant and Editor: **Keith Pledger**

Keith Pledger, Gareth Cole, Joe Petran and Linda Mason

Series Editor: **Roger Porkess**

HODDER
EDUCATION
AN HACHETTE UK COMPANY

Photo credits

p.7 © scanrail – 123RF; p.129t © zuzule – Fotolia; p.129b © Syda Productions – Fotolia; p.130 © Stanislav Halcin – Fotolia; p.142 © Ingram Publishing – Thinkstock/Getty Images; p.143 © Lana Langlois – 123RF.

Although every effort has been made to ensure that website addresses are correct at time of going to press, Hodder Education cannot be held responsible for the content of any website mentioned. It is sometimes possible to find a relocated web page by typing in the address of the home page for a website in the URL window of your browser.

Orders: please contact Bookpoint Ltd, 130 Milton Park, Abingdon, Oxon OX14 4SB. Telephone: (44) 01235 827720. Fax: (44) 01235 400454. Lines are open 9.00–17.00, Monday to Saturday, with a 24-hour message answering service. Visit our website at www.hoddereducation.co.uk.

© Keith Pledger, Gareth Cole, Joe Petran, Linda Mason 2016

First published in 2016 by

Hodder Education

An Hachette UK Company,

50 Victoria Embankment

London EC4Y 0DZ

Impression number	5	4	3	2
Year	2020	2019	2018	2017

Cover photo © ShpilbergStudios

Illustrations by Integra

Typeset in India by Integra Software Services Pvt. Ltd., Pondicherry

Printed in Great Britain by CPI Group (UK) Ltd, Croydon CR0 4YY

A catalogue record for this title is available from the British Library.

ISBN 978 1471 874604

Contents

■ Units with this symbol are required for the Mathematics GCSE only.

NUMBER

Strand 2 Using our number system

Strand 3 Accuracy

Strand 4 Fractions

Strand 5 Percentages

Strand 6 Ratio and proportion

Strand 7 Number properties

ALGEBRA

GEOMETRY AND MEASURES

STATISTICS AND PROBABILITY

How to get the most from this book

Introduction

This book is part of the Mastering Mathematics for WJEC GCSE series and supports the textbook by providing lots of extra practice questions for the Intermediate tier in Mathematics and Mathematics – Numeracy.

This Practice Book is structured to match the Intermediate Student's Book and is likewise organised by key areas of the specification: Number, Algebra, Geometry & Measures and Statistics & Probability. Every chapter in this book accompanies its corresponding chapter from the textbook, with matching titles for ease of use.

Please note: the 'Moving On' units in the Student's Book cover prior knowledge only, so do not have accompanying chapters in this Practice Book. For this reason, although the running order of the Practice Book follows the Student's Book, you may notice that some Strand/Unit numbers appear to be missing, or do not start at '1'.

Progression through each chapter

Chapters include a range of questions that increase in difficulty as you progress through the exercise. There are three levels of difficulty across the Student's Books and Practice Books in this series. These are denoted by shaded spots on the right hand side of each page.

Low difficulty

Medium difficulty

High difficulty

You might wish to start at the beginning of each chapter and work through so you can see how you are progressing.

Question types

There is also a range of question types included in each chapter, which are denoted by codes to the left hand side of the question or sub-question where they appear. These are examples of the types of question that you will need to practice in readiness for the GCSE Intermediate Maths exam.

PS Practising skills

These questions are all about building and mastering the essential techniques that you need to succeed.

DF Developing fluency

These give you practice of using your skills for a variety of purposes and contexts, building your confidence to tackle any type of question.

PB Problem solving

These give practice of using your problem solving skills in order to tackle more demanding problems in the real world, in other subjects and within Maths itself.

Next to any question, including the above question types, you may also see the below code. This means that it is an exam-style question.

ES Exam style

This question reflects the language, style and wording of a question that you might see in your GCSE Intermediate Maths exam.

Answers

There are answers to every question within the book on our website.

Please visit: www.hoddereducation.co.uk/MasteringmathsforWJECGCSE

Number Strand 2 Using our number system Unit 5 Using the number system effectively

PS ― PRACTISING SKILLS   DF ― DEVELOPING FLUENCY   PB ― PROBLEM SOLVING   ES ― EXAM-STYLE

PS 1 Work these out.

  a 8400 × 1000   **b** 8400 × 100   **c** 8400 × 10

  d 8400 × 0.1   **e** 8400 × 0.01   **f** 8400 × 0.001

  g 4978 × 1000   **h** 4978 × 100   **i** 4978 × 10

  j 4978 × 0.1   **k** 4978 × 0.01   **l** 4978 × 0.001

PS 2 Work these out.

  a 60 × 0.1   **b** 340 × 0.1   **c** 5400 × 0.01

  d 2230 × 0.01   **e** 690 × 0.001   **f** 223 × 0.001

PS 3 Work these out.

  a 2.1 × 0.1   **b** 6.25 × 0.1   **c** 13.7 × 0.01

  d 245.6 × 0.01   **e** 0.3 × 0.001   **f** 4.57 × 0.001

PS 4 Work these out.

  a 2.1 ÷ 0.1   **b** 6.25 ÷ 0.1   **c** 13.7 ÷ 0.01

  d 245.6 ÷ 0.01   **e** 0.3 ÷ 0.001   **f** 4.57 ÷ 0.001

DF 5 Write the answers to these calculations in order of size, smallest first.

  a 4.8 × 0.1   **b** 3.56 ÷ 0.1   **c** 29.8 × 0.01

  d 75.5 ÷ 0.01   **e** 19.9 × 0.001   **f** 0.72 ÷ 0.001

DF **6** Find the missing numbers. ●○○

    a $250 \times \boxed{} = 25$ **b** $250 \div \boxed{} = 25$ **c** $1.98 \times \boxed{} = 1980$

    d $1.98 \div \boxed{} = 1980$ **e** $654 \times \boxed{} = 6.54$ **f** $654 \div \boxed{} = 6.54$

PB **7** A game of mathematical 'snap' uses cards as shown below. ●○○
Use arrows to show which two cards are equal.

8.88×0.1	88.8×0.001	$8.88 \div 0.1$
a	**b**	**c**

$8.88 \div 0.01$	888×0.1	$888 \div 0.1$
d	**e**	**f**

PB **8** The signpost in Mathsland offers you four
routes to your destination.
Which is the shortest? ●○○

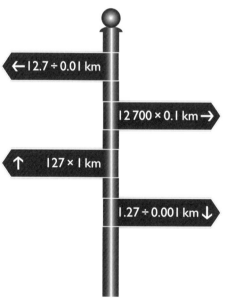

← $12.7 \div 0.01$ km

$12\,700 \times 0.1$ km →

↑ 127×1 km

$1.27 \div 0.001$ km ↓

PB **9** Here are some calculations involving numbers. ●○○

2.56×100	$2.56 \div 100$	2.56×0.1

$2.56 \div 0.001$	2.56×0.001	$2.56 \div 0.1$

Arrange them in order of size from largest to smallest.

Number Strand 2 Using our number system Unit 6 Understanding standard form

PS — PRACTISING SKILLS   DF — DEVELOPING FLUENCY   PB — PROBLEM SOLVING   ES — EXAM-STYLE

PS 1 Write these numbers in standard form.

   a 847

   b 84 700

   c 0.000 847

   d 0.000 000 847

PS 2 Write these numbers in standard form.

   a 620

   b 820 000

   c 20 million

   d 1 millionth

PS 3 Write these as ordinary numbers.

   a 8.52×10^2

   b 3.4×10^{-3}

   c 2.02×10^5

   d 5.762×10^8

   e 4.55×10^{-7}

PS 4 Write these numbers in standard form.

   a 0.003 45

   b 0.000 005 48

   c 0.000 765 4

   d 0.000 000 234 5

DF **5** Write these numbers in standard form.

    a Eight thousand

    b Four fifths

    c Six hundredths

DF **6** Write these quantities in standard form.

    a The distance between the Earth and the Sun is 93 million miles.

    b The area of one person's skin is about $15\,000\,cm^2$.

    c The distance from the equator to the north pole is $20\,000\,km$.

    d There are about 400 million stars in the Milky Way.

    e The area of the UK is $243\,610\,km^2$.

DF **7** Write these numbers in order, starting with the smallest.

    a 4.2×10^{-3}

    b 7.21×10^{-2}

    c 0.09

    d 8.2×10^{-3}

    e 5.7×10^{2}

    f 3.6×10^{3}

    g 6.2×10^{2}

    h 0.57

PB **8** The table shows the distances of the Sun to nearby stars.

An astronomer is researching the stars and needs to make sure his list is in order, with the stars nearest to the Sun first. Show the list he needs to write.

Star	Distance from the Sun (in km)
Procyon B	1.08×10^{14}
Barnard's Star	5.67×10^{13}
Proxima Centauri	3.97×10^{13}
Sirius A	8.136×10^{13}
Ross 128	1.031×10^{14}
Wolf 359	7.285×10^{13}
Rigil Kentaurus	4.07×10^{13}
Luyten 726	7.95×10^{13}

Number Strand 2 Using our number system Unit 7 Calculating with standard form

PS — PRACTISING SKILLS   DF — DEVELOPING FLUENCY   PB — PROBLEM SOLVING   ES — EXAM-STYLE

PS **1** Copy and complete each of the following. Replace each letter with the missing number.

a $5.85 \times 10^5 + 2.35 \times 10^5 = a \times 10^5$

b $1.97 \times 10^{-3} + 2.8 \times 10^{-3} = b \times 10^{-3}$

c $7.09 \times 10^7 - 6.3 \times 10^7 = c \times 10^7$

d $9.4 \times 10^{-5} + 9.4 \times 10^{-5} = d \times 10^{-4}$

PS **2** Work out the following, giving your answers in standard form.

a $100 \times 1.8 \times 10^6$

b $1000 \times 9.3 \times 10^7$

c $10000 \times 2.7 \times 10^{-2}$

d $5.3 \times 10^7 \div 1000$

e $1.03 \times 10^3 \div 10000$

f $1.2 \times 10^{-4} \div 100$

DF **3** Given that $x = 3.5 \times 10^5$, $y = 1.8 \times 10^2$ and $z = 2 \times 10^{-3}$, work out the value of the following. Give your answers in standard form.

a xy

b $\frac{x}{z}$

c x^2

d z^3

e xyz

f x^{-3}

DF **4** Use the information in the table to answer the following. Give your answers in standard form.

| 1 kilowatt = 10^3 watts |
| 1 megawatt = 10^6 watts |
| 1 gigawatt = 10^9 watts |
| 1 terawatt = 10^{12} watts |

a Change 230 gigawatts to watts.

b Change 0.25 gigawatts to kilowatts.

c Change 125 kilowatts to megawatts.

d Change 18 500 megawatts to terawatts.

DF **5** The mass of a blue whale is 1.9×10^5 kg.

The mass of a house mouse is 1.9×10^{-2} kg.

How many times greater than the mass of the house mouse is the mass of the blue whale?

DF **6** Scientists estimate:

- there are about 100 billion galaxies in the observable universe and
- each galaxy contains an average of 300 billion stars.

Work out an estimate for the total number of stars in the observable universe.

Give your answer in standard form.

(1 billion $= 10^9$)

DF ES **7** The table gives information about the number of litres of water used by a factory for seven days.

Monday	Tuesday	Wednesday	Thursday	Friday	Saturday	Sunday
9.32×10^5	9.85×10^5	1.02×10^6	9.93×10^5	1.18×10^6	1.05×10^6	9.66×10^5

Work out the mean amount of water the factory uses each day.

Give your answer in litres, in standard form.

DF **8** $x = 4.5 \times 10^4$

For each of the following, give your answer in standard form correct to four decimal places.

a Work out

i x^3 **ii** $\sqrt[3]{x}$ **iii** $\dfrac{1}{x}$.

b What number is half way between x and $\sqrt{x}$?

PB ES **9** The speed of light is approximately 3×10^8 m/s and the distance of the Earth from the Sun is approximately 1.5×10^{11} m.

Approximately how many seconds does it take for light to travel from the Sun to the Earth?

PB ES **10** The diagram shows a circle drawn inside a square.

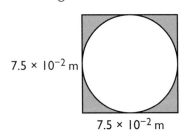

7.5×10^{-2} m

7.5×10^{-2} m

Work out the area, in m², of the shaded part.

Give your answer in standard form correct to 3 significant figures.

Number Strand 3 Accuracy
Unit 5 Approximating

PS — PRACTISING SKILLS   DF — DEVELOPING FLUENCY   PB — PROBLEM SOLVING   ES — EXAM-STYLE

PS **1** Work out an estimate for the answers to these calculations.

    a $8.87 + 98.35 - 5.08$

    b 6.9×9.2

    c $210.7 \div 6.89$

    d 18.9^2

    e $\sqrt{99.54}$

    f $\dfrac{29.7 \times 81.5}{0.529}$

PS **2** A packet of batteries costs £1.95.
Which of these amounts gives a sensible
estimate for the cost of five packets of
these batteries?

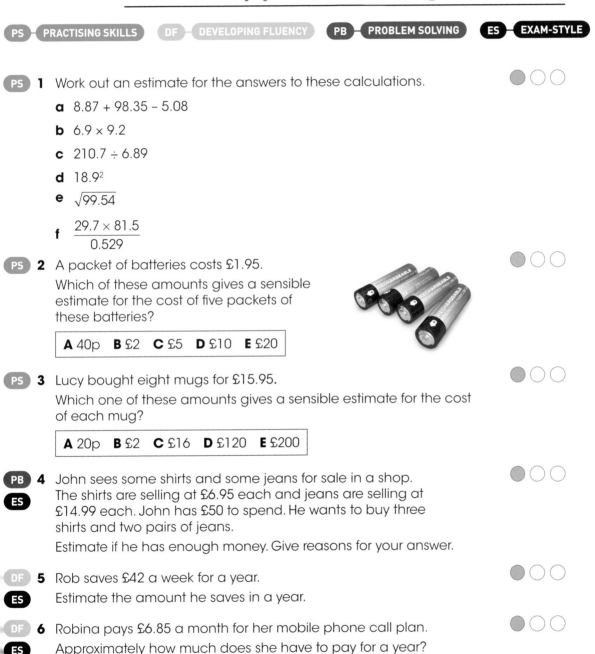

| **A** 40p | **B** £2 | **C** £5 | **D** £10 | **E** £20 |

PS **3** Lucy bought eight mugs for £15.95.
Which one of these amounts gives a sensible estimate for the cost
of each mug?

| **A** 20p | **B** £2 | **C** £16 | **D** £120 | **E** £200 |

PB **ES** **4** John sees some shirts and some jeans for sale in a shop.
The shirts are selling at £6.95 each and jeans are selling at
£14.99 each. John has £50 to spend. He wants to buy three
shirts and two pairs of jeans.
Estimate if he has enough money. Give reasons for your answer.

DF **ES** **5** Rob saves £42 a week for a year.
Estimate the amount he saves in a year.

DF **ES** **6** Robina pays £6.85 a month for her mobile phone call plan.
Approximately how much does she have to pay for a year?

PB
ES
7 Here is part of Harry's electricity bill. It shows the units he used during May.

Harry has saved £65 to pay this bill. Estimate if he has saved enough. Explain your answer.

Electricity 4us	
Mr H Styles	June 4th
12 Lower Rd	
End of May reading	5396
End of April reading	4979
Units used	417
Cost per unit	9.5 pence
Monthly charge	£19.50

DF **8** Becky drove a distance of 295 miles in 5 hours 14 minutes.
Make an estimate of her average speed in miles per hour.

DF **9** Liam saves 5% of his pay each month for five years. His average monthly pay is £1025.
Estimate how much money Liam has saved in the 5 years.

DF **10** It takes Rachel two hours 10 minutes a day to travel to and from work. She works 5 days a week for 48 weeks a year.
Estimate how many hours Rachel spends travelling in a year.

PB **11** Rodney buys items at car boot sales and sells them on an internet website. Here are the items he bought and sold last week.
Find an approximation for Rodney's total profit last week.

Item	Bought for	Sold for
Doll	£9.99	£21.65
Jigsaw	£5.05	£2.95
Camera	£15.75	£39.85
Computer game	£4.99	£21.06
China ornament	£5.00	£99.99

PB
ES
12 Airi owns a factory that makes plastic toys. The factory makes 395 toys an hour. 10% of the toys are not perfect. The perfect toys are put into bags. The bags are then packed into cartons. Each carton holds 48 bags. The factory makes toys for 8 hours 5 minutes each day.
Estimate the number of cartons that are needed each day.
You must show all your working.

Number Strand 3 Accuracy
Unit 6 Significance

PS — PRACTISING SKILLS   DF — DEVELOPING FLUENCY   PB — PROBLEM SOLVING   ES — EXAM-STYLE

PS **1** Write down the number of significant figures there are in each of these numbers.

   a 2.75

   b 507

   c 0.0045

   d 1009

   e 0.0306

   f 1.0

PS **2** Write these numbers to two significant figures.

   a 2.75

   b 507

   c 0.00453

   d 1009

   e 0.0306

   f 1.02

PS **3** Write the number 2 367 450 correct to

   a three significant figures

   b one significant figure

   c two significant figures.

PS **4** Write the number 0.399 99 correct to

   a three significant figures

   b one significant figure

   c two significant figures.

PB **5** 17 845 people attended an outdoor concert in a park.

 a The local newspaper gave the attendance to three significant figures. Write the attendance to three significant figures.

 b The local Radio station gave the attendance to two significant figures. Write the attendance to two significant figures.

 c The local TV reporter gave the attendance to one significant figure. Write the attendance to one significant figure.

PB **6** The formula for working out the area of an ellipse is πab.

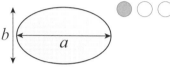

 a Using the value of π as 3.141 59, find the area of the ellipse where $a = 4.50$ and $b = 2.76$. Give the answer correct to three significant figures.

 b Find the difference in the answer to part **a** and the answer if the value of π being used was written to three significant figures. Give the answer correct to three significant figures.

PB **7** Here is part of Gerri's gas bill. It shows the units she used during March.

 Work out Gerri's gas bill for March. Give your answer correct to two significant figures.

Gas 2U	
Mrs G Hall	April 4th
2 High St	
End of March reading	4593
End of February reading	3976
Units used	617
Cost per unit	15.6 pence
Monthly charge	£10.50

PB **ES** **8** The Earth is 92 955 807 miles away from the Sun. The speed of light is 186 000 miles per second correct to 3 significant figures.

 Find the number of seconds it takes for a ray of light to leave the Sun and reach the Earth. Give your answer correct to two significant figures.

DF **ES** **9** The area of a square is 8 cm².

 Find the length of one side of the square. Give your answer correct to three siginificant figures.

8 cm²

DF **ES** **10** The volume of a cube is 10 cm³.

 Find the length of one side of the cube. Give your answer correct to three significant figures.

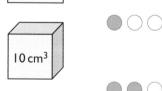

10 cm³

PS **ES** **11** Paula ran 10.55 miles in 1 hour 54 minutes.

 Work out her average speed in miles per hour. Give your answer correct to two significant figures.

Number Strand 3 Accuracy
Unit 7 Limits of accuracy

PS — PRACTISING SKILLS   DF — DEVELOPING FLUENCY   PB — PROBLEM SOLVING   ES — EXAM-STYLE

PS **1** Write down the lower and upper bounds for each of these measurements.

   a 2300 m (to the nearest m)

   b 2300 m (to the nearest 10 m)

   c 2300 m (to the nearest 50 m)

PS **2** Each of these measurements is rounded to the number of decimal places given in brackets.
Write down the lower and upper bounds for each measurement.

   a 7.8 ml (to 1 decimal place)

   b 0.3 ml (to 1 decimal place)

   c 0.31 m (to 2 decimal places)

   d 0.058 m (to 3 decimal places)

PS **3** Each of these measurements is rounded to the given number of significant figures.
Write down the lower and upper bounds for each measurement.

   a 9 g (to 1 significant figure)

   b 90 g (to 1 significant figure)

   c 84 cm (to 2 significant figures)

   d 0.84 cm (to 2 significant figures)

PS **4** Copy and complete the inequality statement for each part.

   a The length of a ladder is x cm. To the nearest 10 cm, the length is 370 cm.

      ☐ $\leq x <$ ☐

   b The mass of an egg is m g. To the nearest gram, the mass is 57 g.

      ☐ $\leq m <$ ☐

   c The body temperature of a baby is T °C. To 1 decimal place, the temperature is 36.4 °C.

      ☐ $\leq T <$ ☐

   d The capacity of a saucepan is y litres. To 2 significant figures, the capacity is 2.8 litres.

      ☐ $\leq y <$ ☐

DF **5** $x = 56.7$ (to 1 decimal place) and $y = 84.2$ (to 1 decimal place).

    a Work out the lower bound for x.

    b Work out the lower bound for y.

    c Work out the lower bound for $x + y$.

DF **6** A square field has side length 65 m, to the nearest metre.
    Work out the lower and upper bounds for

    a the perimeter of the square

    b the area of the square.

DF **7** Given that $23.5 \leqslant l < 24.5$ and $17.5 \leqslant m < 18.5$, work out the upper
    bound for $l - m$.

PB **ES** **8** A stadium sells premium tickets and standard tickets.
    The cost of a premium ticket is £25.00.
    The cost of a standard ticket is £12.50.
    On Saturday:

    • 2500 people buy a premium ticket (to the nearest 100)

    • 7400 people buy a standard ticket (to the nearest 100).

    Let T be the total amount of money paid for premium tickets and standard
    tickets.
    Work out the lower bound and the upper bound for T.

PB **ES** **9** Alis recorded the time taken, to the nearest 10 seconds, for a
    cashier to serve each of four customers in a shop.
    Here are her results.
    150   220   190   110
    Work out the lower bound for the mean time taken to serve
    these customers.

PB **10** The diagram shows a badge that is in the shape of a sector of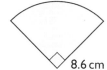
    a circle.
    The radius of the sector of the circle is 8.6 cm (to 2 significant
    figures).
    Work out

    a the upper bound for the perimeter of the badge

8.6 cm

    b the lower bound for the area of the badge.

Number Strand 3 Accuracy
Unit 8 Upper and lower bounds in addition and subtraction

PS — PRACTISING SKILLS   **DF** — DEVELOPING FLUENCY   **PB** — PROBLEM SOLVING   **ES** — EXAM-STYLE

PS **1** Write down the number that is halfway between:

  a 5 and 6

  b 6.5 and 6.6

  c 17.67 and 17.68

  d 2.362 and 2.363

  e 10 and 10.0001

PS **2** Write down the lower bound and the upper bound for these measurements.

  a The length of a pencil is 14 cm to the nearest centimetre.

  b The length of a race is 100 m measured to the nearest centimetre.

  c The weight of a chocolate bar is 75 g to the nearest gram.

  d The weight of a bag of compost is 25 kg to the nearest 100 grams.

  e The capacity of a bottle of milk is 1 litre measured to the nearest 10 ml.

DF **3** Raphael paints pictures. He charges £150 per square metre for every painting he sells. He paints a rectangular picture that has a length of 1.2 m and a width of 80 cm. Both measurements are correct to the nearest centimetre.
Work out the upper and lower bounds of the cost of this picture.

DF **4** The circumference of the Earth around the equator is 24 900 miles correct to the nearest 10 miles.

  a Work out the upper and lower bounds of the diameter of the Earth.

  b What assumption have you made in carrying out this calculation?

DF **5** Rhodri has a ladder that is 10 m long measured correct to the nearest 2 cm.
The base of the ladder has to be 3 m measured to the nearest 5 cm from the base of a wall.
Find the upper bound and the lower bound of the height the ladder can reach up the wall.

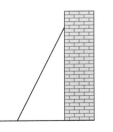

13

PB
ES
6 Peter cycled to work. His average speed was 4.8 m/s correct to 1 decimal place. It took him 20 minutes correct to the nearest minute.

   a Calculate the lower bound of the distance that Peter travelled to work.

Peter took a different route home from work. He cycled a distance of 6.2 km correct to 1 decimal place. It took him 19 minutes correct to the nearest minute.

   b Calculate the upper bound of Peter's average speed in m/s for his journey home.

PB
ES
7 The average fuel consumption (f) of a car, in kilometres per litre, is given by the formula $f = \dfrac{d}{u}$ where d is the distance travelled in kilometres and u is the fuel used in litres.

Jill travels 430 km and uses 52.3 litres of fuel. The 430 is written correct to 3 significant figures. The 52.3 is written correct to one decimal place.

Work out the value of f to a suitable degree of accuracy. You must show all of your working and give a reason for your final answer.

DF
8 Carys lays 50 squares tiles edge to edge in a straight line.
Each tile has a side of length 5 cm, correct to the nearest 2 mm.

   a What is the least length, in cm, of the straight line of these 50 tiles?

   b What is the greatest length, in cm, of the straight line of these 50 tiles?

PB
9 A bucket holds 5 litres of water, correct to the nearest 0.5 litres.
A tank holds 100 litres, correct to the nearest 4 litres.

How many of these buckets of water would it take to be **certain** to be able to fill this tank?

You must show all your working.

PB
10 Dafydd cycles 44 km, correct to the nearest 2 km.

It takes him 3 hours, correct to the nearest $\dfrac{1}{2}$ hour.

   a Calculate Dafydd's greatest average speed, in km/hr for this cycle ride.

   b Calculate Dafydd's least average speed, in km/hr for this cycle ride.

Number Strand 4 Fractions
Unit 6 Dividing fractions

PS — PRACTISING SKILLS   DF — DEVELOPING FLUENCY   PB — PROBLEM SOLVING   ES — EXAM-STYLE

The questions in this unit should be answered *without* the use of a calculator.

 1 Pair each number with its reciprocal.

| $\dfrac{1}{2}$ | 5 | $\dfrac{3}{10}$ | $\dfrac{2}{5}$ | $3\dfrac{1}{3}$ |

| $\dfrac{4}{15}$ | 2 | $2\dfrac{1}{2}$ | $\dfrac{1}{5}$ | $3\dfrac{3}{4}$ |

 2 Change each of these into a multiplication and then work out the answer.

a $5 \div \dfrac{1}{2}$ b $8 \div \dfrac{1}{3}$ c $9 \div \dfrac{1}{4}$

d $\dfrac{3}{5} \div 4$ e $\dfrac{2}{3} \div 5$ f $\dfrac{4}{7} \div 8$

 3 Change each of these into a multiplication and then work out the answer. Cancel the fractions before multiplying.

a $\dfrac{7}{10} \div \dfrac{1}{5}$ b $\dfrac{3}{8} \div \dfrac{1}{4}$ c $\dfrac{5}{9} \div \dfrac{1}{3}$

d $\dfrac{5}{9} \div \dfrac{1}{6}$ e $\dfrac{7}{10} \div \dfrac{14}{15}$ f $\dfrac{15}{24} \div \dfrac{9}{16}$

 4 Change each of these into a multiplication and then work out the answer. Cancel the fractions before multiplying.

a $2\dfrac{1}{7} \div \dfrac{1}{7}$ b $5\dfrac{2}{3} \div \dfrac{1}{6}$ c $1\dfrac{5}{8} \div \dfrac{1}{4}$

d $3\dfrac{1}{9} \div 1\dfrac{1}{6}$ e $4\dfrac{9}{10} \div 1\dfrac{2}{5}$ f $7\dfrac{4}{7} \div 2\dfrac{2}{21}$

DF **5** A bag contains $1\dfrac{3}{5}$ lbs of sugar. A teaspoon holds $\dfrac{1}{150}$ lbs of sugar.

How many teaspoons of sugar are there in the bag?

DF **6** Write the answers to these in order of size. Start with the lowest number. ●●○

$$5\frac{1}{3} \div 1\frac{5}{6}$$　　$$4\frac{2}{5} \div 1\frac{1}{10}$$　　$$7\frac{7}{8} \div 2\frac{3}{4}$$　　$$10\frac{5}{12} \div 3\frac{1}{6}$$

DF **7** Complete this multiplication grid. ●●○

×	$1\frac{2}{5}$	**b**
$3\frac{3}{4}$	**a**	$2\frac{1}{4}$
c	**d**	$5\frac{2}{3}$

PB
ES

8 The diagram shows a rectangular wall.

The area of the wall is $5\frac{5}{9}$ m².

The width of the wall is $1\frac{2}{3}$ m.

Work out the perimeter of the wall.

 ●●○

PB
ES

9 Ravi is going to put some lawn feed on his lawn.

The total area of Ravi's lawn is 125 m².

A packet of lawn feed covers $4\frac{5}{7}$ m² of lawn.

Each packet of lawn feed costs £1.89.

Ravi thinks he can cover his lawn with lawn feed for less than £50.

Is Ravi right? Show how you get your answer.

●●○

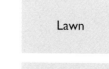

Number Strand 5 Percentages Unit 4 Applying percentage increases and decreases to amounts

PS – PRACTISING SKILLS   **DF** – DEVELOPING FLUENCY   **PB** – PROBLEM SOLVING   **ES** – EXAM-STYLE

PS **1** Work out these.
- **a** **i** 1% of £250
  - **ii** 12% of £250
  - **iii** Decrease £250 by 12%.
- **b** **i** 1% of £21.50
  - **ii** 18% of £21.50
  - **iii** Decrease £21.50 by 18%.

PS **2** Work out these.
- **a** Decrease 350 g by 17%.
- **b** Increase 326 m by 21%.
- **c** Decrease £24.50 by 6%.
- **d** Increase 560 l by 12.5%.
- **e** Decrease 125 cm by 7.5%.
- **f** Increase $1250 by 3.5%.

PS **3** Kim buys a power drill for £68.50 plus VAT at 20%.
How much does she pay?

PS **4** Jose buys a plate for £24. He sells it the next day for a profit of 45%.
How much does Jose sell the plate for?

DF **5** Complete these. Which symbol, <, > or = goes in each box?
- **a** £350 increased by 10% ☐ £430 decreased by 10%
- **b** $49.50 decreased by 15% ☐ $52.40 decreased by 20%
- **c** €128 increased by 29% ☐ €132 increased by 26%
- **d** ₳1250 decreased by 15.8% ☐ ₳1100 increased by 7.5%

DF **6** Ravi travels to work by train. On Monday it took him 1 hour 40 minutes to travel to work. On Tuesday it took him 15% less time to travel to work.

How long did it take Ravi to travel to work on Tuesday?

DF **7** Yasmin has a meal at a restaurant. Here is her bill.

```
        Alf's Chip Shop
  Fish and chips       £9.85
  Tea                  £1.35
  Subtotal a
  15% service charge b
  Total to pay c
```

Work out the missing entries in the bill.

PB **ES** **8** The table gives information about the population of Riverton in 2005 and in 2015.

Marco says: 'The population of Riverton has increased by 10% from 2005 to 2015.' Is he right? Explain your answer.

Year	2005	2015
Population	15310	16678

PB **ES** **9** Fiona works in a warehouse. For the first 20 hours she works in a week she is paid at an hourly rate of £7.80. For each additional hour she works her hourly rate is increased by 35%. Last week Fiona worked 28 hours.

Work out her pay.

PB **ES** **10** Karl measures the lifetimes of two batteries, battery A and battery B. The lifetime of battery A was 36 hours. The lifetime of battery B was 48 hours. The manufacturer of the batteries claims that battery B lasts at least 30% longer than battery A.

Is the manufacturer correct? Explain your answer.

DF **ES** **11** Nerys invests £4800 at 2.4% per annum simple interest. Work out the total value of the investment after 3 years.

PB **ES** **12** In the triangle ABC, angle CAB = 40° and angle ABC is 65% larger than angle CAB.

Work out angle BCA.

PB **13** The diagram shows two circles, circle P and circle Q. The area of circle P is 50 cm². The area of circle Q is 27.5% larger than the area of circle P.

Work out the radius of circle Q. Give your answer correct to 2 decimal places.

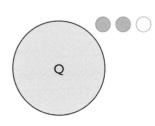

Number Strand 5 Percentages
Unit 5 Finding the percentage change from one amount to another

PS - PRACTISING SKILLS   DF - DEVELOPING FLUENCY   PB - PROBLEM SOLVING   ES - EXAM-STYLE

PS **1** Work out these.

    a Write £2.75 as a percentage of £20.

    b Write 45° as a percentage of 360°.

    c Write 49.5 s as a percentage of 120 s.

PS **2** Write the first number as a percentage of the second number. Give each answer correct to 1 decimal place.

    a 2, 7   **b** 359, 511   **c** 511, 359

    d 18, 10.6   **e** 0.789, 1.249

DF **3** Liam weighed 85 kg at the beginning of his diet and 77 kg at the end of his diet.

    a How much weight did he lose?

    b What is the percentage loss in his weight? Give your answer correct to 1 decimal place.

DF **4** The height of a tree at the beginning of the year is 17.5 m. The height of the tree at the end of the year is 18.9 m.

    Work out the percentage increase in the height of the tree.

PB **ES** **5** The diagram shows information about the counters in a bag.

    a Work out the percentage of blue counters in the bag.

    The percentage of black counters in the bag is greater than the percentage of green counters in the bag.

    b How much greater?

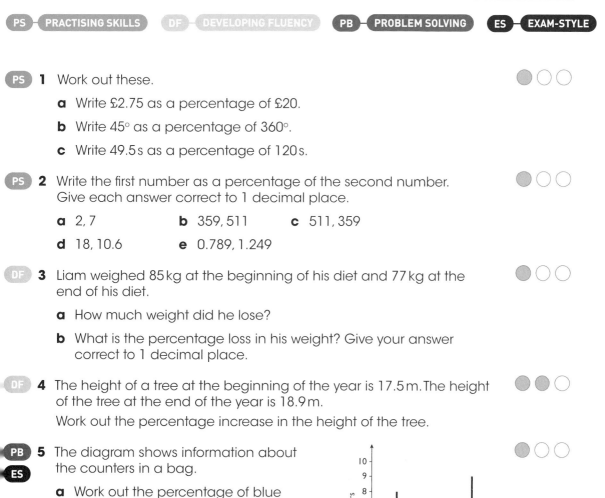

19

PB **ES** **6** A village collects £13 109 to build a new playground.
The target is £20 000. The village magazine says: 'We've collected over 65% of our target, well done!'

Is the village magazine right?
Explain your answer.

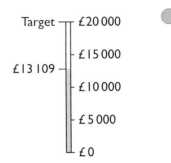

PB **ES** **7** The diagram gives information about the finances of a company.

a What percentage of the company's finances is taxes?

The company director gets a bonus if the profit is 10% greater than the costs.

b Does the company director get a bonus? Give a reason for your answer.

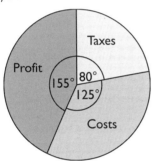

PB **ES** **8** A box contains only arrow shapes and star shapes.

a What percentage of the shapes are star shapes?

Ben is going to add some more arrow shapes to the box. He wants the percentage of arrow shapes in the box to equal 70%.

b How many more arrow shapes does he need to add to the box?

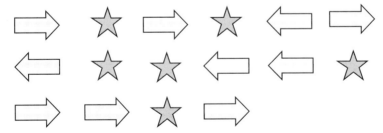

Number Strand 5 Percentages
Unit 6 Reverse percentages

PS — PRACTISING SKILLS   **DF** — DEVELOPING FLUENCY   **PB** — PROBLEM SOLVING   **ES** — EXAM-STYLE

PS **1** A washing machine costs £270 which includes VAT at 20%.
Work out the cost of the washing machine without VAT. ●●○

PS **2** A computer costs £450 following a reduction of 25%.
Work out the cost of the computer before the reduction. ●●○

PS **3** 8717 people visited a tourist attraction in June. ●●○
This is a 15% increase in the number of people who visited the attraction in May.
How many people visited the attraction in May?

PS **4** The cost of a barrel of oil at Toby's garage is £45. This is 60% less than ●●○
it was 5 years ago.
Work out the cost of a barrel of oil at Toby's garage 5 years ago.

PS **5** A special bottle of Mega Juice contains 1.625 litres of orange juice. ●●○
This is 30% more than a standard bottle of Mega Juice.
Work out the amount of orange juice in a standard bottle of Mega Juice.

PS **6** The length of an iron rail at 30 °C is 556.2 cm. This is 3% greater than ●●○
the length of the iron rail at 10 °C.
Work out the length of the iron rail at 10 °C.

PS **7** The installation of a contactless ticket machine at a cinema reduces ●●○
the average time taken to buy a film ticket by 28%. The average time
taken to buy a film ticket using the contactless ticket machine is 153 seconds.
What was the average time taken to buy a film ticket before the
installation of the contactless ticket machine?

DF **8** A sofa costs £840 which includes VAT at 20%. ●●○
Work out the VAT.

DF **9** Work out the original value for each of these. ●●○
   a ☐ cm is increased by 25% to give 107.5 cm.
   b ☐ g is decreased by 5% to give 461.51 g.
   c £ ☐ is increased by 36.5% to give £352.17.
   d ☐ litres is decreased by 17.5% to give 80.85 litres.
   e ☐ km is increased by 0.75% to give 39.091 km.

DF **10** Marc buys a suit, a shirt and a tie in a department store sale.

The price of the suit was reduced by 25% to £81.

The price of the shirt was reduced by 20% to £24.

The price of the tie was reduced by 75% to £5.

a What was the price of the suit, the shirt and the tie before the sale?

b How much money did Marc save?

PB **11** After a dry summer, Bryn Reservoir contained 1.5×10^{10} litres of
ES   water. This is 36% less than the maximum capacity of the reservoir.

Work out the maximum capacity of Bryn Reservoir.

Give your answer in standard form.

PB **12** Llinos worked for 20 hours in her part-time job this week.
ES   This was an increase of $33\frac{1}{3}$% in the number of hours she worked last week.

a How many hours did Llinos work last week?

b This week Llinos's pay is £144.80. This was more than her pay last week.
How much more?

PB **13** The handbook of a motorbike states that the pressure, in pounds per
ES   square inch (psi), of the back tyre of the motorbike should be:

· in the range 40.5–45 psi

· 12.5% greater than the pressure of the front tyre of the motorbike.

Work out the range of possible pressures for the front tyre of the motorbike.

DF **14** Poppy recorded the times taken by some students to complete a
Sudoku puzzle. Here are her results for the male students.

12 min 18 sec   15 min 25 sec   14 min 5 sec   18 min 43 sec   16 min 55 sec

17 min 47 sec   14 min 50 sec   13 min 29 sec   15 min 18 sec   16 min 22 sec

The mean time taken by the male students is 4% less than the mean time taken
by the female students.

Work out the mean time taken by the female students.

PB **15** There are three heptagons, A, B and C.
ES   The area of C is 30% greater than the area of B.

The area of B is 20% greater than the area of A.

The area of C is 70.2 cm².

Work out the area of A.

PB **16** 100 people watched the first round of a darts competition.

122 people watched the second round of the darts competition.

The number of males watching the second round was 20% greater than the
number of males watching the first round.

The number of females watching the second round was 25% greater
than the number of females watching the first round.

Work out the number of female spectators watching the second round
of the competition.

Number Strand 5 Percentages
Unit 7 Repeated percentage increase / decrease

PS PRACTISING SKILLS   DF DEVELOPING FLUENCY   PB PROBLEM SOLVING   ES EXAM-STYLE

PS 1 Write down the meaning of each calculation. The first one has been done for you. ●○○

   a 280 × 1.2   *Increase 280 by 20%.*

   b 280 × 1.25

   c 280 × 1.02

   d 280 × 1.025

   e 280 × 0.8

PS 2 Copy and complete the calculation to work out each percentage change. ●○○

   a Increase 34.5 by 10%.   34.5 × ▢

   b Decrease 304 by 12%.   304 × ▢

   c Decrease 3.125 by 12.5%.   3.125 × ▢

   d Decrease 0.758 by 6.5%.   0.758 × ▢

PS 3 Calculate each of these. ●●○

   a Increase 400 by 10%. Increase the result by 10%.

   b i Increase 520 by 10%. Decrease the result by 10%.

     ii Explain why the answer is not 520.

   c Decrease 1200 by 15%. Increase the result by 20%.

PS 4 Ella invests £5000 into a bank account for 3 years. The bank pays compound interest at an annual rate of 5%. ●●○

   Which calculation represents the value of the investment after 3 years?

   a $5000 \times (0.05)^3$ **b** $5000 \times 0.05 \times 3$

   c $5000 \times (1.05)^3$ **d** $5000 \times 1.05 \times 3$

DF 5 Here is a number machine.

   Input → × 1.15 → × 0.875 → Output

a Copy and complete the table for this number machine.

Input	520	0.8	108.8	2116
Output				

b What does the number machine do to an input number? Give your answer in terms of percentages.

DF **6** The table gives some information about the number of seals on an island each April from 2010 to 2013.

Year	April 2010	April 2011	April 2012	April 2013
Percentage change	No information for previous year	10% more than previous year	7.5% less than previous year	5% more than previous year
Number of seals	8000			

Copy and complete the table.

DF **ES** **7** Mair invests £4000 for 3 years. The investment pays compound interest at an annual rate of 2%.

Harry invests £3800 for 3 years. His investment pays compound interest at an annual rate of 3%.

The total amount of interest that Harry gets for his investment is more than the total amount of interest that Mair gets for her investment.

How much more?

PB **ES** **8** In a sale, the price of handbags are reduced by 30%.

Sam buys a handbag in the sale and uses her loyalty card which gives her a further 10% discount on all items.

The original cost of the handbag is £84.

She pays for the handbag with three £20 notes.

How much change should she get?

PB **ES** **9** The value of a new car depreciates with time.

At the end of the first year, the value of the car is 20% less than its value at the beginning of the year.

At the end of the second year, the value of the car is 15% less than its value at the beginning of the year

At the end of the third year, the value of the car is 10% less than its value at the beginning of the year.

The value of a new car is £16 450.

Work out the value of the car after three years.

Give your answer to the nearest £100.

PB **ES** **10** Clio plants a tree that is 2 m in height.

The height of the tree increases by 10% each year.

How many years will it take for the tree to reach a height of 4 m?

Number Strand 6 Ratio and proportion Unit 2 Sharing in a given ratio

PS PRACTISING SKILLS   **DF** DEVELOPING FLUENCY   **PB** PROBLEM SOLVING   **ES** EXAM-STYLE

PS **1** Work these out.

    a Share 72 in the ratio 1:5.

    b Share 48 in the ratio 1:3.

    c Share 81 in the ratio 2:1.

    d Share 200 in the ratio 4:1.

PS **2** Work these out.

    a Share 60 in the ratio 2:3.

    b Share 91 in the ratio 3:4.

    c Share 105 in the ratio 5:2.

    d Share 250 in the ratio 7:3.

PS **3** Work these out.

    a Share 54 in the ratio 1:3:2.

    b Share 90 in the ratio 5:2:3.

    c Share 117 in the ratio 2:3:4.

    d Share 425 in the ratio 1:1:3.

DF **4** Orange drink is made from orange concentrate and water in the ratio 1:14. Shelly makes some orange drink. She uses 25 ml of orange concentrate.

How much water does she need?

DF **5** A box of chocolates contains milk chocolates and plain chocolates in the ratio 3:4.

What fraction of the box of chocolates are

    a milk chocolates

    b plain chocolates?

DF **6** Haan and Ben win first prize in a tennis doubles competition. The first prize is £600. They share the prize in the ratio 7 : 5.

    a How much does Haan get?

    Haan now shares his part of the prize with Tania in the ratio 2 : 3.

    b How much does Tania get?

DF **7** Emma and Joe share the cost of a meal in the ratio 2 : 5. The cost of the meal is £66.50. Joe pays more than Emma.

    How much more?

DF **8** A bag contains 20p coins and 50p coins in the ratio 7 : 5. There are a total of 180 coins in the bag.

    Work out the total amount of money in the bag.

DF **9** The area of pentagon A and the area of pentagon B are in the ratio 5 : 9. The area of pentagon A is 105 cm².

    Work out the area of pentagon B.

A 105 cm²   B

PB **10** A box contains red, blue and green counters in the ratio 2 : 4 : 3.

ES   **a** What fraction of the counters are blue?

    b What fraction of the counters are not red?

    There are 27 green counters in the bag.

    c Work out the total number of counters in the bag.

PB **11** Viki buys wooden chairs and plastic chairs in the ratio 3 : 7. The cost of each wooden chair is £31.60. The cost of each plastic chair is £15.80.

ES The total cost of the wooden chairs is £284.40.

    How much does Viki pay in total for the plastic chairs?

PB **12** The angles in a triangle are in the ratio 2 : 3 : 7.

ES Show that the triangle is not a right-angled triangle.

PB **13** A box contains only blue pens and black pens in the ratio 5 : 4. Olivia takes 8 blue pens from the box. The number of blue pens in the box is now equal to the number of black pens in the box.

    Work out the total number of pens in the box.

PB **14** The length and the width of a rectangle are in the ratio 5 : 3. The perimeter of the rectangle is 120 cm.

    Work out the area of the rectangle.

Number Strand 6 Ratio and proportion Unit 3 Working with proportional quantities

PS — PRACTISING SKILLS   DF — DEVELOPING FLUENCY   PB — PROBLEM SOLVING   ES — EXAM-STYLE

PS 1 Seven batteries cost a total of £8.75.

    a How much does 1 battery cost?

    b How much do 5 batteries cost?

PS 2 Eight calculators cost a total of £46.80.

    a How much do 5 calculators cost?

    b How much do 13 calculators cost?

PS 3 There are 180 packets of crisps in 5 boxes. How many packets of crisps are there in

    a 3 boxes   **b** 8 boxes?

DF 4 Here is a recipe to make 12 almond shortbread biscuits.
Grandma is going to use this recipe to make 21 biscuits.
How much does she need of each ingredient?

```
          Almond shortbread biscuits
(makes 12 biscuits)
5 oz butter              8 oz flour
1 oz ground almonds      3 oz caster sugar
```

DF 5 The label on a 0.75 litre bottle of Fruit Squash says it makes 60 drinks.
What should the label on a 1.75 litre bottle of Fruit Squash say about the number of drinks it makes?

DF 6 Boxes of paperclips come in two sizes and prices.

    a For the small box of paperclips, work out the cost of 1 paperclip.

    b Which box is the better value for money? Explain your answer.

Large box

Small box

50 paperclips
£1.40

225 paperclips
£6.75

DF **7** A spring stretches 6.3 cm when a force of 28 newtons (28 N) is applied to it.

    a How much will the spring stretch when a force of 15 N is applied to it?

The spring stretches 2.7 cm when a force *F*N is applied to it.

    b Work out the value of *F*.

PB **8** The table gives information about Mani's pay for last week.
ES   This week Mani worked 30 hours at a standard rate and 10 hours at a bonus rate.

How much more did he earn this week compared with last week?

	Number of hours worked	Total
Standard rate	35	£273.70
Bonus rate	5	£60.80
		£334.50

PB **9** Aabish is going to make some concrete. She has 100 kg of cement,
ES   180 kg of sharp sand, 400 kg of aggregate and an unlimited supply of water.

Work out the greatest amount of concrete Aabish can make.

Materials for concrete (makes 0.125 m³)	
Cement	40 kg
Sharp sand	75 kg
Aggregate	150 kg
Water	22 *l*

PB **10** Baked beans come in three sizes of can. The table gives information
ES   about these cans.

Which size of can is the better value for money? Explain your answer.

Size of can	Weight of baked beans (grams)	Cost (p)
Small	180	28
Medium	415	64
Large	840	130

PB **11** The height of the Statue of Liberty is 305 feet. The height of St. Paul's
      Cathedral is 111 metres. (10 feet is approximately 3 metres)

Which is taller, the Statue of Liberty or St. Paul's Cathedral?

Number Strand 7 Number properties Unit 4 Index notation

PS — PRACTISING SKILLS   **DF** — DEVELOPING FLUENCY   **PB** — PROBLEM SOLVING   **ES** — EXAM-STYLE

PS **1** Write each of these as a power of 2.

   a $2 \times 2 \times 2$

   b $2 \times 2 \times 2 \times 2 \times 2 \times 2 \times 2 \times 2 \times 2$

   c $(2 \times 2 \times 2 \times 2) \times (2 \times 2)$

   d $(2 \times 2) \times (2 \times 2) \times (2 \times 2)$

PB **ES** **2** Change $1\,m^2$ to mm^2.

   Give your answer as a single power of 10.

PS **3** Write each of these as a power of 3.

   a $\dfrac{3 \times 3 \times 3}{3}$

   b $\dfrac{3 \times 3 \times 3 \times 3 \times 3}{3 \times 3}$

   c $\dfrac{3 \times 3 \times 3 \times 3 \times 3 \times 3 \times 3}{3 \times 3 \times 3}$

   d $\dfrac{(3 \times 3 \times 3 \times 3 \times 3) \times (3 \times 3 \times 3)}{3 \times (3 \times 3 \times 3)}$

PS **4** Write each of these in index form.

   a $(2 \times 2 \times 2 \times 2) \div (2 \times 2)$

   b $(5 \times 5 \times 5 \times 5 \times 5) \div (5 \times 5 \times 5)$

   c $(7 \times 7 \times 7 \times 7 \times 7) \div (7 \times 7)$

   d $(11 \times 11 \times 11) \div (11 \times 11)$

PS **5** Write each of these as a single power of 3.

   a $3^2 \times 3^3$

   b $3 \times 3^2 \times 3^3$

   c $27 \times 9 \times 81$

PS **6** Write each of these as a single power of 5.

   a $5^4 \div 5^2$

   b $5^4 \div 5$

   c $5^5 \div 125$

PS **7** Write each of these in index form.

   a $2 \times 2 \times 2 \times 3 \times 3$

   b $2 \times 3 \times 2 \times 2 \times 2 \times 3$

   c $5 \times 7 \times 7 \times 7 \times 7 \times 7 \times 5$

   d $3 \times 3 \times 2 \times 5 \times 2 \times 5 \times 2 \times 5$

DF **8** Write each of these as a single power of 2.

   a 4^2 **b** 8^2 **c** 16^2

DF **9** Work out the area of each shape. Give your answers in index form.

a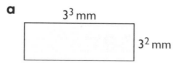

b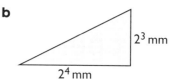

c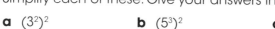

d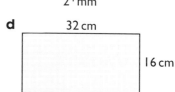

DF **10** Simplify each of these. Give your answers in index form.

    a $(3^2)^2$ **b** $(5^3)^2$ **c** $(2^2)^3$ **d** $(11^4)^2$

DF **11** Simplify each of these. Give your answers in index form.

    a $\dfrac{3^7}{3^2 \times 3^3}$ **b** $\dfrac{7^5 \times 7^3}{7^3 \times 7^2}$ **c** $\dfrac{5^3 \times 5^2}{5^3 \div 5^2}$ **d** $\dfrac{(2^4)^2}{2^3}$ **e** $\dfrac{(3^3)^3}{(3^2)^3}$

PB **ES** **12** There are 64 squares on a chessboard.

Milo is going to put 1 grain of barley on the first square of the chessboard, 2 grains on the second square, 4 grains on the third square, 8 grains on the fourth square, and so on.

    a How many grains of barley should Milo put on the 64th square of the chessboard?
        Give your answer in the form 2^n, where n is a number to be found.

    b 16 grains of barley weigh 1 g. Work out the total mass of the barley on the 64th square of the chessboard.
        Give your answer in kilograms, in standard form and correct to 3 significant figures.

PB **ES** **13 a** Write 125 as a power of 5.

    b Show that $125^2 = 25^3$.

PB **ES** **14** You are given that

    $x = 3 \times 5^3$

    $y = 2 \times 5^2$ and

    $x + 5y = 5^n$, where n is an integer.

    Work out the value of n.

PB **15** The formula $P = 2^n - 1$, where n is a prime number, is used to find prime numbers (P).

    Can the formula be used to find all possible prime numbers?
    Give a reason for your answer.

Number Strand 7 Number properties Unit 5 Prime factorisation

PS — PRACTISING SKILLS   DF — DEVELOPING FLUENCY   PB — PROBLEM SOLVING   ES — EXAM-STYLE

DF **1** Here are some numbers given as products of their prime factors. Write the numbers in order of size, starting with the smallest.

2×7^2 $2^2 \times 3 \times 5$ $2 \times 3 \times 11$ $2^3 \times 3^2$

PS **2** Write each of these in index form.

    a $3 \times 3 \times 2$ **b** $5 \times 3 \times 5 \times 5$ **c** $5 \times 7 \times 5 \times 7$

    d $2 \times 5 \times 3 \times 5 \times 2$ **e** $3 \times 2 \times 3 \times 2 \times 3$ **f** $2 \times 7 \times 2 \times 2 \times 7$

PS **3** Write each number as a product of its prime factors in index form.

    a 105   **b** 165   **c** 315   **d** 150

PS **4** Copy and complete the table.

Numbers	Factors	Common factors	Highest common factor, HCF
15, 20	15: 1, 3, 5, 15 20: 1, 2, 4, 5, 10, 20	1, 5	5
8, 28	8: 1, 2, 4, 8 28: 1, 2, 4, 7, 14, 28	1, 2, 4	
16, 40	16: 1, 2, 4, 8, 16 40: 1, 2, 4, 5, 8, 10, 20, 40		
24, 36			

PS **5** Find the HCF of each pair of numbers.

    a 18 and 30   **b** 27 and 36   **c** 28 and 70   **d** 52 and 130

PS **6** Write down the LCM of each pair.

    a 90 and 135   **b** 15 and 50   **c** 24 and 40   **d** 18 and 48

DF **7** Copy and complete these statements.

    a $2 \times 5 \times 7^2 = \boxed{}$ **b** $3 \times 5^2 \times \boxed{} = 825$ **c** $2 \times 3^{\boxed{}} \times 5 = 90$

    d $2 \times 3 \times \boxed{}^2 = 294$ **e** $2^3 \times \boxed{}^2 = 200$ **f** $2^2 \times 3^{\boxed{}} \times 5^2 = 2700$

DF **8** $30 = 2 \times 3 \times 5$ and $105 = 3 \times 5 \times 7$

a Which prime factors are common to 30 and 105?

b Copy and complete the Venn diagram.

c Write down the HCF of 30 and 105.

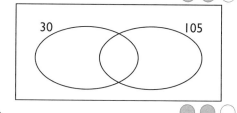

DF **9** **a** Write 126 as a product of its prime factors.

b Write 180 as a product of its prime factors.

c Draw a Venn diagram to show the prime factors of 126 and 180.

d Write down the HCF of 126 and 180.

PB **ES** **10** The table shows the quantities and costs of pencils and erasers.

	Pencils	Erasers
Number in a box	8	10
Cost	£1.80	£2.40

Simon wants to buy as many pencils and erasers as possible, but he wants to buy exactly the same number of each. He has £40.

Work out the total cost of pencils and erasers that Simon should buy.

PB **ES** **11** The diagram shows three lighthouses, A, B and C.

The lights on lighthouse A flash once every 5 seconds.

The lights on lighthouse B flash once every 7 seconds.

The lights on lighthouse C flash once every 15 seconds.

Peter is standing on an island watching the lighthouses.

At 21:00, all three lighthouses flash together.

How many times will all three lighthouses flash together during the next hour?

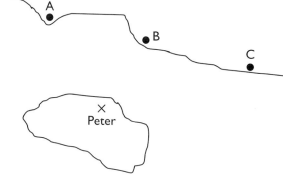

PB **ES** **12** The diagram shows two cogs, X and Y.

An arrow is drawn on each cog.

Cog X has 10 teeth.

Cog Y has 12 teeth.

Cog X is turned in a clockwise direction until the arrows return to their starting positions.

Work out the angle through which cog X is turned.

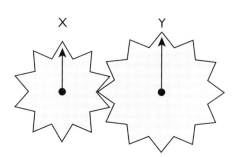

Number Strand 7 Number properties Unit 6 Rules of indices

PS PRACTISING SKILLS   DF DEVELOPING FLUENCY   PB PROBLEM SOLVING   ES EXAM-STYLE

PS **1** Copy and complete the table by writing each number as a single power of 5.

Ordinary number	125	25	5	1	$\frac{1}{5}$	$\frac{1}{25}$	$\frac{1}{125}$
Index form		5^2					

PS **2** Use the rule $a^n \times a^m = a^{n+m}$ to simplify each of these. Give your answers in index form.

   a $2^3 \times 2^4$

   b $2^{-1} \times 2^5$

   c $7^3 \times 7^0$

PS **3** Use the rule $a^n \div a^m = a^{n-m}$ to simplify each of these. Give your answers in index form.

   a $5^7 \div 5^6$

   b $7^{-1} \div 7^{-1}$

   c $\frac{11^{-2}}{11^8}$

PS **4** Use the rule $(a^m)^n = a^{m \times n}$ to simplify each of these. Give your answers in index form.

   a $(5^2)^3$

   b $(2^5)^0$

   c $(11^{-3})^{-2}$

PS **5** Write each of these as a fraction in its simplest form.
   For example, $2^{-3} = \frac{1}{2^3} = \frac{1}{8}$.

   a 2^{-2} **b** 3^{-2} **c** 11^{-1} **d** 10^{-3}

DF **6** Without using a calculator, state which of these is equal to 1.

   a 5^0 **b** $3^2 \times 3^{-2}$ **c** $(5^3)^{-2}$ **d** $2^5 \div 2^5$ **e** $(7^3)^0$

DF **7** Write each of these as a single power of 2.

    a 4   **b** 4^2 **c** $(4^3)^2$ **d** $(4^5)^4$

DF **8** Work out each of these. Give your answers in index form.

    a $(2^2 \times 2^3) \times 2^4$

    b $(7^3 \times 7^4) \div 7^5$

    c $(2^3)^2 \times (2^2)^3$

    d $(5^4 \div 5^6)^{-1}$

DF **9** $3^4 = 81$ and $3^5 = 243$

    Write the answer to each of these as a single power of 3. Do not use a calculator.

    a 9×81

    b $\dfrac{243}{9}$

    c $\dfrac{243 \times 81^2}{27}$

DF **10** Use the rules of indices to simplify each of these. Give your answers in index form.

    a $2^3 \times 2^4 \times 3^4 \times 3^2$

    b $\dfrac{3^5 \times 5^4}{3^2 \times 5^2}$

    c $(2^5 \times 5^3)^2$

DF **11** Without using a calculator, copy and complete these statements.
    Use < or > or =.

    a 2^3 ☐ 3^2

    b 2^{-1} ☐ 3^{-1}

    c 3^0 ☐ 3

PB **ES** **12** $5^3 = 125$ and $5^5 = 3125$
    Pierre says $125^{10} = 3125^6$.
    Without using a calculator, say whether or not he is correct. Explain why.

PB **ES** **13** Aled says $(a^m)^n = (a^n)^m$.
    Is he correct? Explain why.

PB **ES** **14** Write down the HCF of each pair of numbers.

    a 42 and 70

    b 42^2 and 70^2

Number Strand 7 Number properties Unit 7 Fractional indices

PS — PRACTISING SKILLS   DF — DEVELOPING FLUENCY   PB — PROBLEM SOLVING   ES — EXAM-STYLE

PS **1** Write the following as roots.

a $5^{\frac{1}{2}}$

b $4^{\frac{1}{3}}$

c $3^{\frac{1}{4}}$

d $8^{\frac{1}{2}}$

e $10^{\frac{1}{5}}$

f $6^{\frac{1}{5}}$

PS **2** Write the following using indices.

a $\sqrt{7}$

b $\sqrt[3]{9}$

c $\sqrt[3]{4}$

d $\sqrt{5}$

e $\sqrt[6]{5}$

f $\sqrt[7]{2}$

PS **3** Find the value of each of these.

a $81^{\frac{1}{2}}$

b $8^{\frac{1}{3}}$

c $256^{\frac{1}{4}}$

d $169^{\frac{1}{2}}$

e $216^{\frac{1}{3}}$

f $625^{\frac{1}{4}}$

DF **4** Write the following using roots.

a $5^{\frac{3}{2}}$

b $7^{\frac{2}{3}}$

c $6^{\frac{3}{4}}$

d $10^{\frac{5}{3}}$

e $10^{\frac{2}{5}}$

f $5^{\frac{5}{2}}$

DF **5** Write the following using indices.

a $\sqrt{2^3}$

b $(\sqrt[4]{3})^3$

c $\sqrt[3]{5^7}$

d $\sqrt[4]{7^5}$

e $(\sqrt[3]{2})^5$

f $(\sqrt[5]{2})^9$

DF **6** Find the value of each of these.

a $36^{\frac{3}{2}}$

b $8^{\frac{2}{3}}$

c $256^{\frac{3}{4}}$

d $4^{\frac{5}{2}}$

e $27^{\frac{5}{3}}$

f $625^{\frac{3}{4}}$

DF **7** Write the following as roots.

a $3^{-\frac{1}{2}}$

b $4^{-\frac{1}{3}}$

c $5^{-\frac{3}{4}}$

d $7^{-\frac{3}{2}}$

e $9^{-\frac{4}{5}}$

f $3^{-\frac{2}{5}}$

DF **8** Write the following using indices.

a $\dfrac{1}{\sqrt{5}}$

b $\dfrac{1}{\sqrt[4]{7}}$

c $\dfrac{1}{\sqrt[3]{5^2}}$

d $\dfrac{1}{(\sqrt{7})^3}$

e $\dfrac{1}{\sqrt[4]{3^5}}$

f $\dfrac{1}{(\sqrt[5]{3})^3}$

DF **9** Find the value of each of these.

a $16^{-\frac{3}{2}}$

b $64^{-\frac{2}{3}}$

c $125^{-\frac{5}{3}}$

d $4^{-\frac{3}{2}}$

e $27^{-\frac{2}{3}}$

f $64^{-\frac{5}{6}}$

PB **10** Write the value of each of these as a power of 2.

a $4 \times 32^{\frac{3}{5}}$

b $\dfrac{1}{8} \times 64^{\frac{3}{2}}$

c $8^{-\frac{5}{3}} \times 32^{\frac{2}{5}}$

PB **11** Find the value of n.

a $\dfrac{1}{\sqrt{8}} = 2^n$

b $\sqrt[3]{27^2} = 3^n$

c $(\sqrt[3]{125})^4 = 25^n$

PB **12** **a** Work out $\left(\dfrac{125}{27}\right)^{-\frac{2}{3}}$

ES **b** Find the value of p in this numeric identity.

$3 \times 8^{\frac{2}{3}} = 96 \times p^{-\frac{1}{3}}$

Algebra Strand 1 Starting algebra Unit 4 Working with formulae

PS — PRACTISING SKILLS   DF — DEVELOPING FLUENCY   PB — PROBLEM SOLVING   ES — EXAM-STYLE

PS 1 Copy and complete the table for this number machine.   ● ○ ○

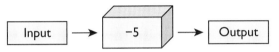

Input → -5 → Output

	Input	Output
a	20	
b		12
c	2	
d		-7
e	n	
f		p

PS 2 Copy and complete the table for this number machine.   ● ○ ○

Input → ×4 → Output

	Input	Output
a	3	
b		56
c	0.5	
d		17
e	m	
f		q

DF **3** Angus thinks of a number. He multiplies the number by 5.
He then adds 3.

    a Work out the result if the number Angus thought of was

       i 2

       ii 10

       iii n.

    b Work out the number Angus thought of when the result was

       i 38

       ii 3

       iii p.

DF **4** Here is a rule to change a quantity in litres into a quantity in pints.

> Multiply the number of litres by 1.75 to get the number of pints.

    a Change 40 litres to pints.

    b Change 1050 gallons to litres. 1 gallon = 8 pints.

PB **ES** **5** Pamela hires a car. The cost £C, for the hire of a car for n days, in two different garages, is shown in the boxes below.

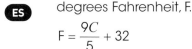

Bill's Autos	**Carmart**
$C = 11n + 60$	$C = 20n$

    a If Pamela hires a car for 8 days, show that Bill's Autos is the cheaper of the two garages.

    b For how many days would Pamela have to hire the car for Carmart to be the cheaper garage?

PB **ES** **6** Here is a formula to change degrees centigrade, C, to degrees Fahrenheit, F.

$$F = \frac{9C}{5} + 32$$

    a On the 1st August, the temperature in New York was 79 °F.
On the 1st August, the temperature in Barcelona was 25 °C.
In which city was the temperature the greater?

    b Show that –20 °C is a higher temperature than –5 °F.

    c Work out the difference between 30 °C and 100 °F.

Algebra Strand 1 Starting algebra Unit 5 Setting up and solving simple equations

PS — **PRACTISING SKILLS**   **DF** — DEVELOPING FLUENCY   **PB** — PROBLEM SOLVING   **ES** — EXAM-STYLE

DF **1** Which of these equations does not have a solution $x = 3$?

  a $x + 5 = 8$ **b** $7 - 2x = 1$ **c** $1 - x = 4$

PB **ES** **2** The lengths of the sides of a rectangle are given by $x + 1$, $3x - 2$, $9 - x$ and $x + 6$.

  Work out the perimeter of the rectangle.

PB **ES** **3** Geoff pays £7.20 for 3 pies and 2 portions of chips.

  Mel pays £4.50 for 5 portions of chips.

  Sally buys 2 pies. How much does Sally pay?

PB **ES** **4** Work out the size of the largest angle.

  $x + 120$ $2x + 50$

  $5x - 10$

PB **ES** **5** Tom, Lucy and Sadiq share the driving on a journey of 145 miles. Tom drives x miles. Lucy drives three times as far as Tom. Sadiq drives 40 miles more than Lucy.

  How many miles do they each drive?

PB **ES** **6** Ceri thinks of a number. She multiplies her number by 2 and then adds 3. Simon thinks of a number. He multiplies his number by 3 and then subtracts 2. Ceri and Simon both think of the same number.

  What is the number they both thought of?

Algebra Strand 1 Starting algebra Unit 6 Using brackets

 PS — PRACTISING SKILLS    DF — DEVELOPING FLUENCY    PB — PROBLEM SOLVING    ES — EXAM-STYLE

PS **1** **a** Expand

    i $4(3m + 5)$

    ii $7(h - 3k)$.

  b Factorise

    i $12x - 8y$

    ii $6z + 6$.

PS **ES** **2** Orange juice costs £x per glass. Cola costs £y per glass. Ham sandwiches cost £a each. Cheese sandwiches cost £b each. Salad sandwiches cost £c each.

  a Sally, James and Hannah each have an orange juice and a cheese sandwich. Brian and Gary each have a cola and a salad sandwich. Write down an algebraic expression for the total cost.

  b Three different people each choose a drink and a sandwich. The total cost is $3y + 2a + c$. Write down what each had.

DF **ES** **3** Write down an expression for

  a the perimeter of this rectangle

  b the area of this rectangle.

Give your answers in their simplest form.

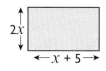
$2x$
$\leftarrow x + 5 \rightarrow$

PS **4** **a** Expand

    i $3(b - 1) + 2(3 - b)$

    ii $5(a + 2) - 3(1 - a)$

  b Factorise

    i $6p^2 - 10p$

    ii $3c^2d + 9cd^2$.

DF
ES

5 Ella and Liam expand $2x(3x - 4)$.

This is what Ella wrote: $2x(3x - 4) = 6x^2 - 4$.

This is what Liam wrote: $2x(3x - 4) = 6x - 8x = -2x$.

a Explain the mistake that each of Ella and Liam made.

b Expand $2x(3x - 4)$.

PB
ES

6 CA, AB and BD are three sides of a quadrilateral of side $(x + 2)$ cm. The fourth side, X, is twice as long as the other three.

Show that the perimeter of the quadrilateral ABCD can be written $5x + 10$.

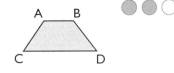

PB
ES

7 Colin is n years old. Della is 4 years older than Colin. Ezra is twice as old as Della.

Show that the sum of their ages is divisible by 4.

PB
ES

8 The cost of hiring a car is £C per day for the first 4 days. The cost is £$(C - 5)$ per day for all additional days. Steve hires a car for 10 days.

a Write down an expression, in terms of C, for the total amount Steve has to pay. Give your answer in its simplest form.

Anne pays £$4(3C - 10)$ for the hire of a car.

b How many days did Anne hire a car for?

PB
ES

9 The diagram shows a path around three sides of a lawn in a garden. The width of the path is x metres. The garden is in the shape of a rectangle of dimensions 20 m by 12 m.

Find, in terms of x, the perimeter of the lawn. Give your answer in its simplest form.

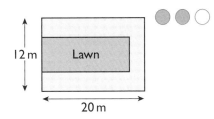

PB
ES

10 Leanne thinks of a whole number. She adds 4 to her number and then multiplies the result by 5. Isaac thinks of a whole number. He subtracts 3 from his number and then multiplies the result by 8.

Show that Leanne and Isaac could not possibly have thought of the same number.

Algebra Strand 1 Starting algebra Unit 7 Working with more complex equations

PS — PRACTISING SKILLS   **DF** — DEVELOPING FLUENCY   **PB** — PROBLEM SOLVING   **ES** — EXAM-STYLE

PS
ES
1 Solve these equations.

    a $5 - 4x = 24$

    b $2a - 7 = 1 - 3a$

PB
ES
2 Mari and Rhodri collect football stickers.

    There are x stickers in a full set of football stickers.

    Mari has 3 full sets and 8 stickers. Rhodri has 1 full set and 32 stickers.

    They each have the same number of stickers.

    How many stickers are in a full set?

PB
ES
3 Bryn pays £7.80 for 3 meat pies and 2 steak puddings.

    Misha pays £12.60 for 7 meat pies.

    Lisa pays £2.45 for 1 meat pie and 1 portion of chips.

    Work out the cost of

    a 1 portion of chips

    b 1 steak pudding.

DF
4 Alex and Josh try to solve this equation.

    $8x - 5 = 2x + 10$

    Alex's solution

    $-5 - 10 = 2x - 8x$

        $+15 = -6x$

           $x = 15 \div -6 = -2.5$

    Josh's solution

    $8x - 2x = 10 + 5$

        $6x = 15$

        $x = 15 - 6 = 9$

    a Both Alex and Josh have made a mistake. Explain each boy's mistake.

    b Find the correct solution.

DF 5 Jenny and Rhea need to solve this equation.

$5y + 13 = 1 - y$

Here is how they start.

Jenny	Rhea
$5y + y = 1 - 13$	$13 - 1 = -y - 5y$
$6y = -12$	$12 = -6y$
$y = \ldots\ldots\ldots$	$y = \ldots\ldots\ldots$

a Complete their solutions.

b Which method do you prefer? Explain why.

PB ES 6 Here is a rectangle. All the measurements are in centimetres.

Work out the perimeter of this rectangle.

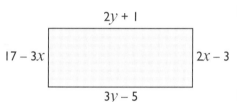

PB ES 7 Tom is x years old.

Bronwyn is 4 times as old as Tom.

Tom is 5 years older than Pat.

In 12 years' time, Bronwyn will be 9 years older than the sum of Tom and Pat's ages.

How old was Pat last year?

PB ES 8 Carriage E and carriage F are two carriages on a train.

At Birmingham, there are 37 people in carriage E and 40 people in carriage F.

At Stoke:

• Three times as many people leave carriage E as leave carriage F.

• 20 people get on the train and go into carriage E.

• 11 people get on the train and go into carriage F.

The number of people in carriage E is now the same as the number of people in carriage F.

How many people from carriage E got off the train at Stoke?

PB ES 9 The diagram shows a square and an equilateral triangle. All measurements are in centimetres.

The perimeter of the square is equal to the perimeter of the equilateral triangle.

Work out the area of the square.

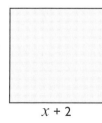
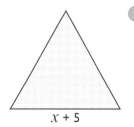

PB ES 10 The angles of an isosceles triangle are $(x - 8)°$, $(5x - 16)°$ and $(144 - 3x)°$.

Work out the size of the smallest angle.

Algebra Strand 1 Starting algebra Unit 8 Solving equations with brackets

PS – PRACTISING SKILLS   **DF** – DEVELOPING FLUENCY   **PB** – PROBLEM SOLVING   **ES** – EXAM-STYLE

PS **ES** **1** Solve these equations.

    a $3 - 2x = 3(x - 7)$

    b $1 - 2(a + 1) = 4(a - 5)$

DF **ES** **2** Toby is trying to solve this equation.

    $4x - 3 = 1 - (x - 4)$

    Here is his solution.

    $4x - 3 = 1 - x - 4$

    $4x + x = 1 - 4 + 3$

       $5x = -6$

        $x = -1.2$

    Toby has made two mistakes.

    Explain these mistakes.

PB **ES** **3** The length of this rectangle is twice the width.

    Work out the perimeter of the rectangle.

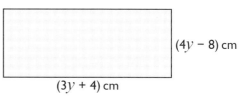

$(4y - 8)$ cm

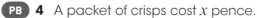

$(3y + 4)$ cm

PB **ES** **4** A packet of crisps cost x pence.

    A can of cola costs 70p.

    Three boys each buy a packet of crisps and a can of cola.

    The total cost is £5.70.

    Arwyn wants to buy two cans of cola and a packet of crisps.

    How much will this cost?

PB **ES** **5** Ruby thinks of a number between 1 and 15.

    When she subtracts the number from 25 and then multiplies the result by 3, she gets the same answer as when she multiplies the number by 4 and then subtracts 9 from the result.

    What number is Ruby thinking of?

PB **ES** **6** Rhian is going to cover a floor with exactly 300 rectangular tiles like the one shown in the diagram.

There will be 15 rows with 20 tiles in each row.

The floor is in the shape of a rectangle with dimensions 5 m by 2.4 m.

What are the dimensions of each tile?

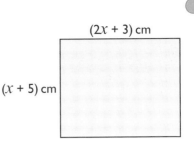

$(2x + 3)$ cm

$(x + 5)$ cm

PB **ES** **7** When Zach was three years old, his height was h cm.

The table shows the increases in Zach's height over the next four years.

Age in years	3	4	5	6	7
Increase in height in cm		5	3	2	2

Zach's father is 1.80 metres tall.

When Zach was seven, he was half the height of his father.

Work out Zach's height when he was three.

PB **ES** **8** In this isosceles triangle, the two equal angles are given by the expression $(x + 20)°$.

Show that the third angle can be written as $2(70 - x)°$.

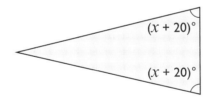

$(x + 20)°$

$(x + 20)°$

PB **ES** **9** The radius of a circle is $(3x - 1)$ cm.

The circumference of the circle is equal to the perimeter of a square of side πx cm.

Show that the area of the circle can be written as 4π cm².

PB **ES** **10** On the diagram, all measurements are in centimetres and all angles are right angles.

Show that the perimeter could never be equal to 32 cm.

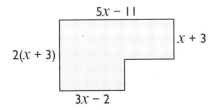

$5x - 11$

$x + 3$

$2(x + 3)$

$3x - 2$

Algebra Strand 1 Starting algebra Unit 9 Simplifying harder expressions

PS — PRACTISING SKILLS   DF — DEVELOPING FLUENCY   PB — PROBLEM SOLVING   ES — EXAM-STYLE

PB **ES** **1** Waqar, Nathan and Wesley play for the school football team.

Waqar has scored 5 more goals than Nathan.

If Nathan scores another goal, he will have scored twice as many goals as Wesley.

Wesley has scored g goals. The three boys have scored a total of T goals.

Write down an expression for T in terms of g.

PB **2** Tom thinks of a number n and adds 4.

Jane thinks of a number m and subtracts 7.

Write down and expand an expression for the product of their results.

PB **ES** **3** Write down an expression, in terms of x, for the area of this shape.

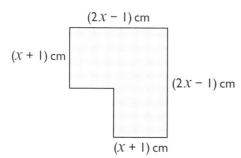

$(2x - 1)$ cm

$(x + 1)$ cm

$(2x - 1)$ cm

$(x + 1)$ cm

PB **ES** **4** The diagram shows a lawn in the shape of a square with a path around it.

The lawn is of side $(x + 3)$ m.

The width of the path is 1 m.

Write down an expression, in terms of x, for the total area of the path.

PB
ES **5** The diagram shows a square and a right-angled triangle.
Show that the area of the square is never equal to the area of the right-angled triangle.

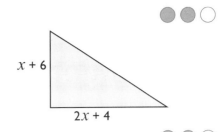

DF
ES **6** Show that the area of this trapezium is $x^2 - 16$.

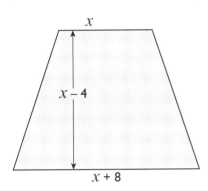

PB
ES **7** Write down in terms of x

a the area of the shaded rectangle

b the area of the shaded triangle.

Expand and simplify your expressions, if necessary.

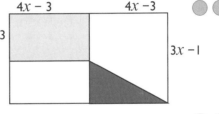

PB
ES **8** Here is a right-angled triangle.
Write down an expression for y in terms of x.

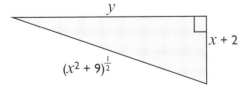

PS
ES **9** **a** Expand and simplify $(y - 5)(y + 8)$.

b Simplify $\dfrac{(2w^2x)^3}{2w^3x \times 3wx^2}$.

DF
ES **10** $\left(x^{n+1}\right)^{n-1} = x^3$

For what value of n is this statement true?

Algebra Strand 1 Starting algebra Unit 10 Using complex formulae

PB ES 1 BB Cars uses the formula $C = 20 + 12.5Gt$ to work out the cost, £C, of renting a car, where G is the group of the car (1, 2, 3 or 4) and t is the number of days for which the car is rented.

Mair paid £370 to rent a car from BB Cars.

Work out one possibility for the number of days and the group of car that Mair rented.

PS ES 2 Here is a formula.

$v = u + at$

  a Work out the value of v when $u = 25$, $a = -10$ and $t = 3.5$.

  b Rearrange the formula to make a the subject.

  c Work out the value of a when $v = 80$, $u = 60$ and $t = 15$.

DF ES 3 The formula for calculating the volume (V) of a sphere is $V = \dfrac{4}{3}\pi r^3$, where r is the radius of the sphere.

  a Work out the volume of a sphere of radius 4.5 cm. Leave your answer in terms of π.

  b Work out the radius of a sphere of volume 200 mm³. Take $\pi = 3.14$.

PB ES 4 A can of cola is in the shape of a cylinder.

The volume of a cylinder (V) is given by the formula $V = \pi r^2 h$, where r is the radius and h is the height of the cylinder.

Peter buys a can that holds 330 ml of cola.

The radius of the can is 3.25 cm.

Work out the height of the can. Take $\pi = 3.14$.

DF ES 5 The surface area (A) of a solid cylinder can be found using the formula $A = 2\pi r^2 + 2\pi rh$, where r is the radius of the cylinder and h is its height.

  a Work out the surface area of a cylinder with radius 7 cm and height 15 cm. Give your answer in terms of π.

  b Work out the height of a cylinder with surface area 20π and radius 2 cm.

PS **ES** **6** Here is a formula.

$v^2 = u^2 + 2as$

    a Work out the value of v when $u = 20$, $a = 10$ and $s = 11.25$.

    b Rearrange the formula to make u the subject.

    c Work out the value of u when $v = 9$, $a = 4$ and $s = 7$.

PB **ES** **7** To change P pounds into E euros, Pete uses the formula $E = 1.36P$.

To change P pounds into D dollars, Pete uses the formula $D = 1.55P$.

    a Write down a formula that Pete could use to change dollars into euros.

    b Pete sees a watch for sale on an American website for 200 dollars.
The same model of watch is for sale in Spain for 175 euros.
In the UK, this model of watch is sold for 130 pounds.
In which currency is the watch cheapest?

DF **ES** **8** The formula $T = 2\pi\sqrt{\dfrac{l}{g}}$ is used to calculate the time period, T, of

a simple pendulum where l is the length in centimetres and g is the
acceleration due to gravity.

    a Work out the value of T when $l = 160$ cm and $g = 10$ m/s^2.
Give your answer in terms of π.

    b For another simple pendulum, $T = \dfrac{2\pi}{7}$.
Work out the length of this simple pendulum when $g = 9.8$ m/s^2.

PS **ES** **9** Here is a formula.

$c = \sqrt{a^2 + b^2}$

    a Make b the subject of the formula.

    b Work out the value of b when $c = 41$ and $a = 40$.

PS **ES** **10** Here is a formula.

$E = mc^2$

Work out the value of

    a E when $m = 2 \times 10^{30}$ and $c = 3 \times 10^8$

    b m when $E = 4.5 \times 10^{28}$ and $c = 3 \times 10^8$.

49

Algebra Strand 2 Sequences
Unit 3 Linear sequences

PS — PRACTISING SKILLS   DF — DEVELOPING FLUENCY   PB — PROBLEM SOLVING   ES — EXAM-STYLE

DF **ES** **1** Here are the first four terms of a sequence.

2   7   12   17

Here are the first four terms of another sequence.

4   7   10   13

The number 7 is in both sequences.

a Find the next two numbers that are in both sequences.

John says the number 202 is in both sequences.

b Is John right?

DF **ES** **2** Here are Pattern number 3 and Pattern number 4 of a sequence of patterns.

Pattern number 3   Pattern number 4

a Draw Pattern number 1 and Pattern number 2.

b Find the missing numbers in the table for this sequence of patterns.

Pattern number	1	2	3	4	5	10
Number of dots			8	11		

3 Here are the 2nd, 4th and 5th terms of a sequence.

___   15   ___   27   33

DF **a** Write down the 1st and 3rd terms of this sequence.

ES **b** Work out the product of the 6th and 7th terms of this sequence.

Ruth says that all the terms in this sequence are odd numbers.

PB **c** Show that Ruth is right.

PB **ES** **4** The anchor on a boat is lowered 3 metres with each turn of a handle. ●○○
The anchor is already 5 metres below the surface of the sea.

    a How many metres below the surface of the sea will the anchor be after n turns of the handle?

The anchor hits the bottom of the sea after 64 turns of the handle.

    b How deep is the sea?

DF **ES** **5** Find the missing numbers in the table for this pattern of stars made with matchsticks. ●○○

Number of stars	1	2	3	4	8	20	n
Number of matches	10	19	28				

PB **ES** **6** A machine makes parts for a mobile phone. The list shows the number of parts that are made at 1 p.m. and at every 5 minutes after 1 p.m. ●○○

240   265   290   315   340

How many parts will be made at 2.30 p.m?

7 Here are the first four terms of a sequence.

150   138   126   114

PS   **a** Write down the next two terms in this sequence. ●○○

PB   **b** In what position is the first negative number in this sequence? ●○○

ES   **c** Show that the nth term of this sequence can be written in the form $6(a + bn)$. ●●○

PB **ES** **8** Here are the first five terms of a sequence.

3   7   11   15   19

    a **i** Write down the next two terms in this sequence. ●○○

       ii Explain how you got your answer to part **i**. ●○○

    b Find the 15th term of this sequence. ●○○

    c Write down, in terms of n, the nth term of this sequence. ●●○

PB **ES** **9** The nth term of sequence A is $2n + 1$. The nth term of sequence B is $4n - 3$.

    a How many of the first 10 numbers in sequence A are prime numbers? ●○○

    b Show that the sum of all corresponding numbers in each sequence is always an even number. ●●○

Algebra Strand 2 Sequences
Unit 4 Special sequences

PS ─ PRACTISING SKILLS   DF ─ DEVELOPING FLUENCY   PB ─ PROBLEM SOLVING   ES ─ EXAM-STYLE

DF 1 Here are the first eight terms of a sequence.

  0   2   2   4   6   10   16   26

   a Describe the rule for working out the terms in this sequence.

   b Johan says: 'All the terms in this sequence must be even numbers.' Explain why Johan is right.

   c What is special about these numbers?

PS 2
ES Here are the first three patterns in a sequence of patterns.

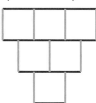

   a Write down the first 5 terms in the sequence formed by the vertical lines.

   b Show that the nth term of this sequence can be written as $\dfrac{n(n+3)}{2}$.

PB 3
ES Here is a sequence of patterns made with dots and straight lines.

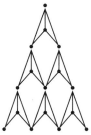

Pattern 1   Pattern 2   Pattern 3

   a Find the missing numbers in the table.

Pattern number	1	2	3	4	10
Number of dots	4	9			
Number of lines	5	15			
Number of triangles	2	6			

b Write down, in terms of n, the nth term for the sequence of dots.
c i Show that the nth term for the sequence of triangles is $n^2 + n$.
  ii Use $n^2 + n$ to help you write, in terms of n, the nth term for the sequence of lines.

 4 Here are the first three terms of a sequence.

  1   3   7

Alan says the next term in this sequence is 13. Becky says the next term in this sequence is 15.

a Explain how **both** Alan and Becky could be right.
b i Write down the 5th term of Alan's sequence.
  ii Write down the 5th term of Becky's sequence.

 5 Here are the first four terms of a sequence.

  16   8   4   2

a Write down the next three terms of this sequence.

b Which of the following expressions is the nth term of this sequence?

$$\frac{32}{2^{n+1}} \quad \frac{n}{2} \quad 2n \quad \frac{32}{2^{n-1}} \quad \frac{32}{2^n}$$

6 P_n is the perimeter of an equilateral triangle.
The next equilateral triangle is formed by joining the midpoints of the triangle.
This is shown in the diagram.

$P_1 = 3s$

Write down P_2, P_3, P_4 and P_5 in terms of s.

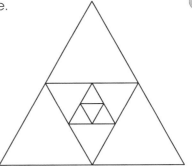

Algebra Strand 2 Sequences
Unit 5 Quadratic sequences

PS — PRACTISING SKILLS   **DF** — DEVELOPING FLUENCY   **PB** — PROBLEM SOLVING   **ES** — EXAM-STYLE

PS **ES** **1** **a** Write down the first four terms of a quadratic sequence with nth term $2n^2 - 6n + 5$.

b Explain why every term of this sequence is an odd number.

PS **ES** **2** Here are the first few terms of a sequence.

0, 4, 18, 48, 100, …

a Explain why this is not a quadratic sequence.

b Find the 10th term.

PB **ES** **3** The nth term of an arithmetic sequence is $2n + 10$.

The nth term of a quadratic sequence is $n^2 - n$.

a Which number appears in both sequences and in the same position?

b Otis says that there are only three terms in the quadratic sequence that are not in the arithmetic sequence. Explain why Otis is correct and write down these three terms.

DF **ES** **4** Here is a pattern made from dots.

Pattern 1   Pattern 2   Pattern 3   Pattern 4   Pattern 5

a Copy the patterns. In each pattern, join each pair of dots with a straight line.

b Write down the number of lines in each pattern.

c The number of lines drawn in each pattern forms a quadratic sequence. Write down the nth term of this sequence.

PB **ES** **5** $n^2 + 4n$ is the nth term of a quadratic sequence.

$2n^2 - 5n$ is the nth term of a different quadratic sequence.

a Angus says that the number 12 appears in both sequences. Is Angus right? Explain your answer.

b For what value of n is the term the same in both sequences?

6 In this pattern, lines are drawn from each vertex to the mid-point of
each side inside some regular polygons.

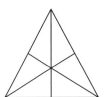 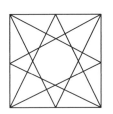 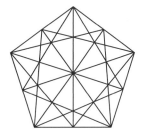

    a Write down the number of lines drawn inside each of these polygons.

    b How many lines would there be inside a regular hexagon?

    c The number of lines in the pattern forms a quadratic sequence. Write down
the nth term for this sequence.

    d How many lines would there be in a 12-sided regular polygon?

 7 Here are five of the first six terms of a quadratic sequence.
    2   4   7   ...   16   22

    a Write down the missing term.

    b What is the 20th term of this sequence?

    c Write down the position-to-term formula.

 8 Here is a pattern made from black and white triangular tiles.

    a Write down an expression in terms of n for the sequence of black triangles.

    b Write down an expression in terms of n for the sequence of white triangles.

    c Show that the sequence formed by the total number of small
triangles in each pattern is the sequence of square numbers.

Algebra Strand 2 Sequences
Unit 6 nth term of quadratic sequences

PS **1** Match each nth term formula to the correct quadratic sequence.

    a $n^2 + 3n$ **A** 6, 8, 8, 6, 2

    b $2n^2 - n - 5$ **B** 0, –1.5, –4, –7.5, –12

    c $5n - n^2 + 2$ **C** –4, 1, 10, 23, 40

    d $\dfrac{1 - n^2}{2}$ **D** 4, 10, 18, 28, 40

PB **ES** **2** Bilal started a computer company in 2010. The profits of the company, in £ millions, for the first five years from 2010 were 0, 2, 6, 12, 20. If this pattern continues

    a what profit might Bilal expect to make in the next year?

    b what profit might Bilal expect to make after n years of the company's existence?

    c in what year will the profits first exceed £100 million?

PB **ES** **3** The nth term of a sequence is given by $n^2 + n + 1$.

    a Kyle says that all the terms in this sequence are prime numbers. Show that Kyle is wrong.

    b How many of the first ten terms are not prime numbers?

PB **ES** **4** Here are the first three patterns in a sequence of patterns made from triangles and hexagons.

  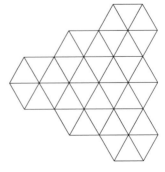

S is the sequence of the number of edges of the hexagons in each pattern. T is the sequence of the number of edges of the small triangles in each pattern.

The sequence formed by subtracting sequence T, term by term, from sequence S has an nth term given by $an^2 + bn$.

Work out the values of a and b.

PB **ES**

5 The nth term of a sequence is given by $n^2 - 2n + 5$. The nth term of a different sequence is given by $n^2 + n - 7$.

   a Which term has the same value and is in the same position in both sequences?

   b Explain why this is the only common term.

PS

6 Work out the nth term of each of these quadratic sequences.

   a $2, 2, 0, -4, -10$

   b $-1, 0, 2, 5, 9$

   c $-6, -4, 0, 6, 14$

   d $-6, -4, 6, 24, 50$

DF **ES**

7 Here are the first three patterns in a sequence of patterns made from sticks.

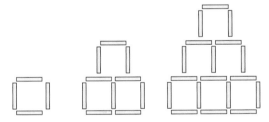

   a Work out the number of sticks in the nth pattern.

   b What is the greatest value of n that can be made with 200 sticks?

PB **ES**

8 Here are the first five terms of a quadratic sequence.

   5   24   55   98   153

The nth term of this sequence is written in the form $an^2 + bn + c$.

   a Work out the values of a, b and c.

   b Prove that all the even numbered terms are even numbers.

Algebra Strand 3 Functions and graphs Unit 2 Plotting graphs of linear functions

PS — PRACTISING SKILLS   DF — DEVELOPING FLUENCY   PB — PROBLEM SOLVING   ES — EXAM-STYLE

DF **1** Milk is sold at 90p per litre.

a Work out the missing values in this table.

Number of litres	1	2	5	10	20	50	100
Cost in £	0.90			9			90

b Draw a graph to show this information about the cost of milk.

c Find the cost of 30 litres of milk.

d How many litres of milk can be bought for £20?

DF **2** Lisa's company pays her travel expenses for each mile she travels.

ES a Work out the missing values in this table.

Miles travelled	5	10	15	20	25	30	35	40
Expenses in £	4	8		16			28	

b Draw a graph to show this information.

c Lisa travels 28 miles. Work out how much her company pays her.

d Lisa's company paid her £60 in travel expenses. How many miles did Lisa travel?

PB **3** This rule can be used to work out the time, in seconds, it takes to download music tracks.

ES

Time = 25 × number of music tracks + 10

a Work out the missing information in this table of values.

Number of tracks	2	4	6		10	12
Time in seconds	60			210		

b Draw a graph to show the time it takes to download music tracks.

c How many tracks can be downloaded in 5 minutes?

PB
ES

4 Vijay lives 3 kilometres from his school. The travel graph shows Vijay's journey to school one day.

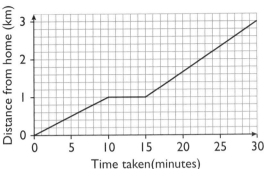

a Describe the three stages in Vijay's journey to school.

On another day, Vijay left home at 08:15. At 08:32 he stopped at a shop to buy a drink. The shop is 1.4 km from school.
At 08:38 he left the shop and carried on walking to school.
He arrived at school at 08:45.

b Draw a travel graph to show this journey.

PB
ES

5 Peter goes on holiday to Prague. The currency in Prague is the Czech Crown (czk). The exchange rate is £1 = 30czk.

a Draw a graph that could be used to convert between £ (pounds) and czk (Czech Crown).

Peter bought a suitcase in Prague. He paid 750czk for the suitcase.
In London the same model of suitcase costs £34.

b How much money did Peter save by buying the suitcase in Prague?

PB
ES

6 Pat delivers parcels. The table shows the cost of delivering parcels for different journeys.

Distance in miles	10	20	30	40	50
Cost in £	20	30	40	50	60

a Draw a graph to show this information.

For each parcel Pat delivers there is a fixed charge plus a charge for each mile.

b Use your graph to work out the fixed charge and the charge for each mile.

Vanessa also delivers parcels. For each parcel Vanessa delivers it costs £1.50 for each mile. There is no fixed charge.

c Compare the cost of having a parcel delivered by Pat with the cost of having a parcel delivered by Vanessa.

DF
ES

7 Draw the graph of $y = 2x + 3$ for values of x from $x = -3$ to $x = 1$.

Algebra Strand 3 Functions and graphs Unit 3 The equation of a straight line

PS — PRACTISING SKILLS   **DF** — DEVELOPING FLUENCY   **PB** — PROBLEM SOLVING   **ES** — EXAM-STYLE

PS **1** Here are some lines drawn on a co-ordinate grid.

    a Write down the equation of

       i the horizontal line passing through A

       ii the vertical line passing through B

       iii the line CD.

    b Draw the lines with equation

       i $x = 2.5$

       ii $y = -3$.

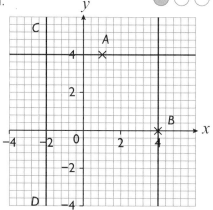

PB
ES **2** The diagram shows the line PQ. A line, L, is parallel to PQ and passes through the point $(9, 0)$. Find an equation of the line L.

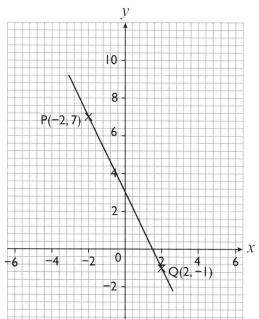

PB **3** Here are some lines drawn on a co-ordinate grid.

a Write down an equation for
  i P
  ii Q
  iii R.

b On a copy of the grid, draw the lines with equations
  i $y = 3x$
  ii $x - y = 3$.

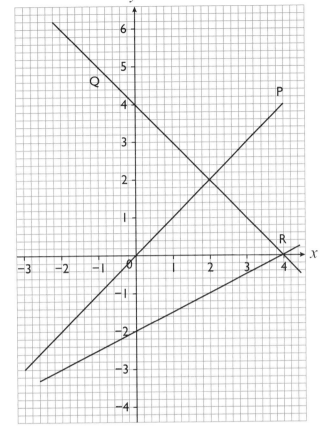

PB **ES** **4** Here are equations of two lines.

$y = 1.5x - 4$ $3x - 2y = 4$

Brian says the two lines are parallel. Jane says the two lines have the same y-intercept.

a Who is right, Brian or Jane, or are they both right?

b Write down the gradient of $y = 1.5x - 4$.

PB **ES** **5** A is a point with co-ordinates $(5, 3)$. B is a point with co-ordinates $(3, -7)$.

a Work out an equation of a straight line passing through A and B.

Another straight line, L, has y intercept of 1. L is parallel to AB.

b Find an equation of L.

PB **ES** **6** The line, M, passes through the point $(-2, 5)$. It is parallel to the line with equation $x + y = 5$.

a Find an equation of the line M.

The line with equation $y = 2x$ intersects M at the point B.

b Work out the co-ordinates of the point B.

c Find the value of x when $y = 8$.

Algebra Strand 3 Functions and graphs Unit 4 Plotting quadratic and cubic graphs

PS — PRACTISING SKILLS   **DF** — DEVELOPING FLUENCY   **PB** — PROBLEM SOLVING   **ES** — EXAM-STYLE

PS **1** **a** This is a table of values for $y = x^2 + 3x - 5$. Work out the missing values in the table.

x	-3	-2	-1	0	1	2	3	4
y	-5			-5		5		

b This is a table of values for $y = 1 + 2x - 3x^2$. Work out the missing values in the table.

x	-3	-2	-1	0	1	2	3	4
y		-15			0			-39

PS **2** **a** This is a table of values for $y = x^3 - x + 4$. Work out the missing values in the table.

x	-3	-2	-1	0	1	2	3	4
y		-2		4	4			64

b This is a table of values for $y = 2x^3 - 6x^2 + 3x - 1$. Work out the missing values in the table.

x	-3	-2	-1	0	1	2	3	4
y			-12		-2		8	

DF **3** The diagram on p. 64 shows part of the graph of $y = 2x^2 - x - 6$.

a Write down the co-ordinates of the y-intercept.

b Write down the solutions of the equation $2x^2 - x - 6 = 0$.

c By considering the line $y = -3$, estimate the solutions of the equation $2x^2 - x - 3 = 0$.

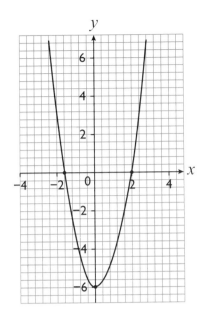

DF **4** The diagram shows part of the graph of $y = x^3 - 2x^2 - 3x$.

    a Write down the solutions of the equation $x^3 - 2x^2 - 3x = 0$.

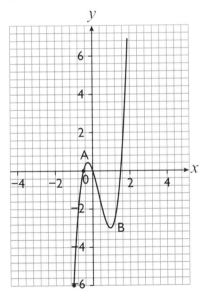

    b Estimate the co-ordinates of the points A and B.

    c Show that the equation $x^3 - 2x^2 - 3x = 4$ has just one solution.

DF **ES** **5** This is a table of values for $y = x^2 - 2x - 3$.

x	–3	–2	–1	0	1	2	3	4
y		5		–3				5

   a Work out the missing values in the table.

   b Draw a graph of $y = x^2 - 2x - 3$.

   c Use your graph to estimate the solutions to $x^2 - 2x - 3 = 0$.

DF **ES** **6** This is a table of values for $y = x^3 - 2x^2 - 5x + 6$.

x	–3	–2	–1	0	1	2	3	4
y				6			0	

   a Work out the missing values in the table.

   b Draw a graph of $y = x^3 - 2x^2 - 5x + 6$.

   c Use your graph to solve $x^3 - 2x^2 - 5x + 6 = 0$.

DF **ES** **7** **a** Draw a graph of $y = 3 + x - 4x^2$.

   b Use your graph to estimate the solutions to $3 + x - 4x^2 = 0$.

DF **ES** **8** On the same axes, draw graphs of $y = x^3 - x$ and $y = x$.
The graphs cross at points A, B and C.

   a Write down the co-ordinates of points A, B and C.

   b The x-co-ordinates of A, B and C are the solutions of a cubic equation. Write down this equation.

   c Write down the equation of a straight line you could draw to solve $x^3 - 3x = 3$.

PB **ES** **9** A cricket ball is thrown vertically up from 2 metres above the ground. Its path is modelled by the quadratic function $h = 2 + 9t - 5t^2$ where h is the height of the ball and t the time.

   a Draw the graph of h against t for values of t from 0 to 2 at intervals of 0.25.

   b Use your graph to find the greatest height the ball reaches from the ground.

MATHEMATICS ONLY

Algebra Strand 3 Functions and graphs Unit 5 Finding equations of straight lines

PS PRACTISING SKILLS   DF DEVELOPING FLUENCY   PB PROBLEM SOLVING   ES EXAM-STYLE

PS 1 Write down the equation of the line that has a gradient of

    a 6 and goes through $(1, 5)$

    b −4 and goes through $(3, 10)$

    c 3.5 and goes through $(-2, -12)$.

DF 2 Triangle PQR is drawn on a co-ordinate grid.

    a Write down the equations of the three lines that make this triangle.

    b Work out the area of triangle PQR.

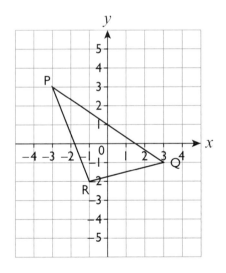

PB **3** The graph shows the prices charged by Safe Car Hire.

ES It shows the relationship between the charge (£C) and the number of days (n) for which the car is hired.

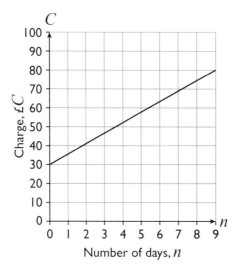

Number of days, n

a i Write down the gradient of this graph.

ii What does this gradient represent?

b Write down the equation of this straight-line graph.

c Alan hires a car from Safe Car Hire for 20 days.
Work out the total charge.

PB **4** The graph shows the time,
T minutes, to cook a turkey
with mass m lb.

ES

T is given by the formula
$T = am + b$.

a Work out the values of
a and b.

b Work out the theoretical
cooking time of a turkey
with a mass of 26 lb.

Give your answer in hours
and minutes.

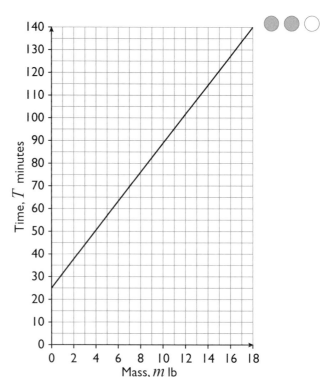

Mass, m lb

66

MATHEMATICS ONLY

PB
ES
5 Line l goes through the points A(0, 5) and B(5, 0).
Line m goes through the points D(0, 2) and C(2, 0).

   a Write down the equation of each line.

   b Work out the area of the quadrilateral ABCD.

DF
ES
6 The two straight lines, p and q, are drawn on a co-ordinate grid.

   a Write down the equation of each line.

   b Write down the equation of the straight line that is parallel to p and passes through (1, 5).

   c What are the co-ordinates of the point where lines p and q intersect?

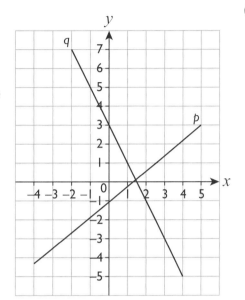

PB
7 A square has vertices A(3, 3), B(3, −2), C(−2, −2) and D(−2, 3).

   a Write down the equations of the diagonals.

   b What is the product of the gradients of the two diagonals?

PS
ES
8 a Write down the gradient of the straight line that joins the points E(5, 8) and F(−3, 20).

   b Toby says that the straight line that goes through E and F will extend through the point G(40, −45).
Is Toby right?

PB
ES
9 Line l passes through points A(7, 2) and B(4, 4).
Line m has equation $2x + 3y = 5$.

   a Prove that l and m are parallel.

   b Line n has gradient 1.5 and passes through the mid-point of AB.
Write down the equation for n.

PB
ES
10 A quadrilateral has vertices A(4, 5), B(9, 2), C(1, −1) and D(−4, 2).

   a Prove that ABCD is a parallelogram.

   b Write down the equation of the line that passes through A and C.

Algebra Strand 3 Functions and graphs Unit 6 Perpendicular lines

PS – PRACTISING SKILLS   DF – DEVELOPING FLUENCY   PB – PROBLEM SOLVING   ES – EXAM-STYLE

PS 1 Look at the diagram.

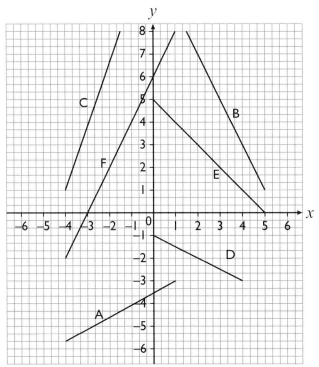

a Which of the lines on the grid are
   i parallel
   ii perpendicular
   to the line with equation $y = 2x - 3$?

b Write down an equation of each of the lines chosen in part **a**.

PS **ES** **2** Write down the gradient of a line perpendicular to the line with equation:

a $y = 2x - 1$

b $y = 1 - 2x$

c $2y = 1 - x$

d $x + 3y = 1$

e $y - 1 = \dfrac{2x}{3}$

f $5x + 4y = 20$

DF **ES** **3** ABC is a right-angled triangle. Angle A = 90°.

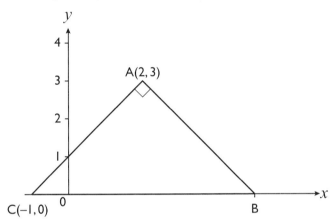

a Write down an equation of the line AB.

b Find the co-ordinates of B.

c Work out the area of triangle ABC.

PS **ES** **4** Here are the equations of eight straight lines.

A $y = 2x - 3$ **B** $y = 5 - 3x$ **C** $3y - x = 7$ **D** $x + y = 3$

E $y = \dfrac{2x - 1}{2}$ **F** $y = 3(1 - x)$ **G** $y = \dfrac{x}{3} - 2$ **H** $x = 3(y + 2)$

a Which lines are parallel?

b Which lines are perpendicular to each other?

PB **ES** **5** Find an equation of a line perpendicular to $y = 4x + 3$ passing through the point (1, 7).

PB **ES** **6** **a** Draw the line with equation $2y = x - 2$. Use the same scale on each axis.

     b Find an equation of the line perpendicular to $2y = x - 2$ passing through $(2, 0)$.

     c Work out the area of the triangle bounded by this perpendicular, the line $2y = x - 2$ and the y-axis.

PB **ES** **7** **a** L is a line with equation $2x + 3y = 6$. On a grid, draw L.

     b The point $(6, -2)$ lies on L. P is a straight line perpendicular to L passing through $(6, -2)$. Find an equation for P.

     c Find the co-ordinates of the intercepts of P with the axes.

     d Find the area bounded by L, P and the x-axis.

PB **ES** **8** AC is a diagonal of a kite ABCD.

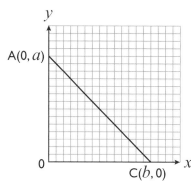

Prove that the equation of the other diagonal can be written as
$$y = \frac{2bx + a^2 - b^2}{2a}.$$

PB **ES** **9** A(2, 2) and B(12, 2) are the end points of a diameter of a semicircle. The point P(4, 6) is a point on the circumference of the semicircle. Find the equation of the tangent to the semicircle at P.

PB **ES** **10** ABCD is a square. A(4, 6) lies on the line with equation $y = 0.5x + 4$. The perpendicular to $y = 0.5x + 4$ passing through A meets the x-axis at B. AB is one side of the square ABCD.

Find the co-ordinates of the points C and D.

Algebra Strand 4 Algebraic methods Unit 1 Trial and improvement

PS ━ PRACTISING SKILLS   **DF** ━ DEVELOPING FLUENCY   **PB** ━ PROBLEM SOLVING   **ES** ━ EXAM-STYLE

DF **1** The equation $x^3 + 2x = 25$ has a solution between 2 and 3. ●●●

    a Calculate this solution correct to 1 decimal place.

    b What test must always be carried out to confirm that the solution is accurate to 1 decimal place?

DF **2** A solution to the equation $2b^3 - b - 42.5 = 0$ lies between 2 and 3. ●●●
Find this solution correct to 2 decimal places.

DF **3** The equation $4x^3 - 50x = 917.7$ has a solution between 6 and 7. ●●●

    a Calculate this solution correct to 1 decimal place.

    b What test must always be carried out to confirm that the solution is accurate to 1 decimal place?

PB **4** The product of $e^2 + 1$ and $e - 1$ is 420.5 ●●●
Use a trial and improvement method to find a positive value of e correct to 1 decimal place.

PB **5** The equation $x(x - 1)(x + 5) = 384$ has a solution between 6 and 7. ●●●

    a Find this solution correct to 2 decimal places.

    b How did your confirm your solution to know that it was accurate to 2 decimal places?

PB **6** The base of a box is a square. ●●●
The height of the box is twice the length of one of the lengths of the base of the box.
The volume of the box is 194.67 cm.
Use a trial and improvement method to find the dimensions of the box, with each measurement correct to 1 decimal place.

PB **7** A triangle has a base of length $(x + 2)$ cm and a perpendicular height ●●●
of $(x - 5)$ cm.
The area of the triangle is 33.5 cm².
Use a trial and improvement method to find the perpendicular height of the triangle correct to 1 decimal place.

Algebra Strand 4 Algebraic methods Unit 2 Linear inequalities

PS ▸ PRACTISING SKILLS   **DF** ▸ DEVELOPING FLUENCY   **PB** ▸ PROBLEM SOLVING   **ES** ▸ EXAM-STYLE

PS **1** Solve these inequalities.

    a $4 - 2x \geqslant 8$

    b $3 + 2x < 5x - 9$

    c $2(x + 3) + 3(2x + 5) \geqslant 37$

PS **ES** **2** **a** Write down the inequality shown on this number line.

    b Show the inequality $-1 \leqslant x < 5$ on a number line.

    c Solve $2x + 3 > 8$.

DF **3** Colin earns £N each year.

Brian earns at least twice as much as Colin.

Becky earns less than half of what Colin earns.

If Brian earns £x each year and Becky earns £y each year, write inequalities to show their earnings in terms of N.

PB **ES** **4** Paris thinks of a number greater than 5.

She subtracts 3 from the number, then doubles the result.

Her final answer is less than 12.

Write down all the possible numbers Paris could have thought of.

PB **ES** **5** The perimeter of this rectangle is at least 44 cm, but less than 50 cm.

Write down an inequality to show the possible values of x.

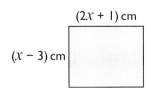

$(2x + 1)$ cm

$(x - 3)$ cm

MATHEMATICS ONLY

PB **ES** **6** April, Bavinda and Chas each have some marbles.
April has 15 more marbles than Bavinda.
Bavinda has three times as many marbles as Chas.
Together they have less than 200 marbles.
What is the greatest number of marbles that April can have?

PB **ES** **7** Write down an inequality for each of the four boundaries of the
shaded region.

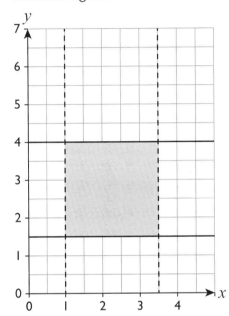

PB **ES** **8** On graph paper, show the region that satisfies the inequalities
$2 \leqslant x < 4$ and $2.5 \leqslant y < 4.5$.

PB **ES** **9** The length, width and height of
this cuboid satisfy the following
inequalities:
$10 \leqslant l < 11 \quad 3 \leqslant w < 3.5 \quad 6 \leqslant h < 8$

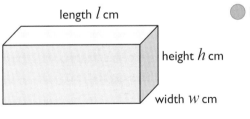

length l cm

height h cm

width w cm

    a Work out the greatest possible
volume of the cuboid.

    b Work out the smallest possible surface area of the cuboid.

PB **ES** **10** n is an integer that satisfies the inequality $5n - 1 > 4n + 2$.
Which one of these inequalities can n not satisfy?
$4n + 5 > n + 2 \qquad 2n - 7 > 1 - n \qquad n + 4 > 6n - 8 \qquad 7 - n > 2n - 11$

PB **ES** **11** n is an integer that satisfies both of these inequalities:
$4n - 1 < 2n + 3$ and $5(n + 4) \geqslant 2(n + 5)$
Write down all the possible values of n.

Algebra Strand 4 Algebraic methods Unit 3 Solving pairs of equations by substitution

PS — PRACTISING SKILLS   **DF** — DEVELOPING FLUENCY   **PB** — PROBLEM SOLVING   **ES** — EXAM-STYLE

PS **1** Solve these pairs of simultaneous equations by substitution.

    a $2x + y = 6$
       $y = x + 3$

    b $x + 4y = 11$
       $x = y + 1$

    c $2x + 3y = 6$
       $x = 5 - 2y$

DF **ES** **2** Aled thinks of two numbers.
    The difference between the two numbers is 7.
    The sum of the two numbers is 25.
    What two numbers is Aled thinking of?

DF **ES** **3** The sum of two numbers is 160.
    The difference between the two numbers is 102.
    Work out the two numbers.

PB **ES** **4** Elfyn pays £10.50 for 4 portions of fish and 3 portions of chips
    Tracey pays £5.40 for 3 portions of fish.
    Malcolm buys 2 portions of fish and 2 portions of chips.
    How much should this cost him?

PB **ES** **5** Chan has exactly 24 notes in his wallet.
    They are either £20 notes or £10 notes.
    The total value of the notes is £410.
    How many £20 notes are in Chan's wallet?

PB **ES** **6** What is the perimeter of this rectangle?

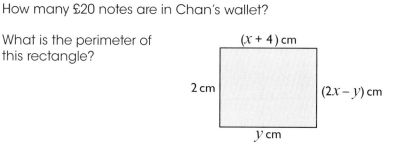

$(x + 4)$ cm

2 cm

$(2x - y)$ cm

y cm

PB **ES** **7** In Year 10:

- There are b boys and g girls.

- There are 45 more boys than girls.

If 12 boys and no girls joined Year 10, there would be twice as many boys as there are girls.

How many students are in Year 10?

PB **ES** **8** A train carriage has t tables that have seats for 4 people and s single seats (with no tables).

The total number of seats in a carriage is 62 and there are 22 more single seats than tables.

In one carriage, 19 of the single seats are occupied.

How many of the single seats are not occupied?

PB **ES** **9** A tennis club has f female members and m male members.

In 2010, they had 16 more female members than male members.

By 2015, the number of female members had increased by one-third, the number of male members had decreased by 18 and there were a total of 120 members.

How many female members did the club have in 2010?

PB **ES** **10** An equation of line l is given by $y = mx + c$.

The points $(1, 7)$ and $(3, 11)$ lie on line l.

Work out the values of m and c.

Algebra Strand 4 Algebraic methods Unit 4 Solving simultaneous equations by elimination

PS – PRACTISING SKILLS   **DF** – DEVELOPING FLUENCY   **PB** – PROBLEM SOLVING   **ES** – EXAM-STYLE

DF **1** Solve these pairs of simultaneous equations by elimination.

    a $3x + 2y = 14$
       $5x - 2y = 18$

    b $2x + 3y = 2$
       $8x + 3y = 17$

    c $6x - 5y = 23$
       $4x - 3y = 14$

PB **ES** **2** The price of tickets for a football match are £a for adults and £c for children.

Morgan pays £270 for tickets for 2 adults and 5 children.

Jim pays £251 for tickets for 3 adults and 2 children.

Peter has £150 to buy tickets for the football match. Does he have enough money to buy tickets for himself and his 3 children?

PB **ES** **3** A taxi company charges a fixed amount plus an additional cost per mile.

A journey of 8 miles costs £8.90. A journey of 12 miles costs £12.10.

Sioned is 20 miles from home. She has only £20.

Does Sioned have enough money to travel home by taxi?

DF **ES** **4** Work out the area of this rectangle.

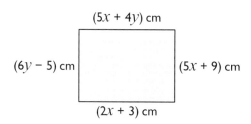

$(5x + 4y)$ cm

$(6y - 5)$ cm

$(5x + 9)$ cm

$(2x + 3)$ cm

MATHEMATICS ONLY

PB **ES** **5** The Smith family and the Jones family have booked the same summer holiday.

Mr and Mrs Smith and their three children paid £2440.

Mr Jones, his mother and father and his son paid £2330.

After they book, the travel company reduces the cost of a child's holiday by 10% and refunds both families.

How much refund should each family receive?

PB **ES** **6** The diagram shows an equilateral triangle and a square.

The perimeter of the square is equal to the perimeter of the triangle.

Work out the area of the square.

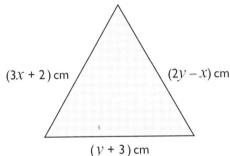 

$(3x + 2)$ cm $(2y - x)$ cm

$(y + 3)$ cm

PB **ES** **7** A boy travels for x hours at a speed of 5 km/h.

He then travels for y hours at a speed of 10 km/h.

In total, he travels 35 km at an average speed of 7 km/h.

Work out the values of x and y.

PB **ES** **8** In one week, Liz works 35 hours at her standard rate of pay and 12 hours at her overtime rate. For this, she is paid £428.40.

In a different week, Liz works 40 hours at her standard rate of pay and 8 hours at her overtime rate. For this, she is paid £425.60.

a Work out Liz's standard rate of pay.

b Work out the ratio of the standard rate of pay to the overtime rate of pay.

PB **ES** **9** Two people bought identical Christmas decorations from the same shop.

One paid £65.60 for 200 streamers and 220 tree decorations.

The other paid £63.30 for 210 streamers and 200 tree decorations.

How much would it cost to buy 200 streamers and 200 tree decorations from this shop?

PB **ES** **10** The points (2, 2.5) and (6, –2.5) lie on the line with equation $ax + by = c$.

a Bob says the point (–2 , 8) also lies on this line.

Is Bob correct?

b Write down

i the gradient of this line

ii the co-ordinates of the intercepts on the axes.

Algebra Strand 4 Algebraic methods Unit 5 Using graphs to solve simultaneous equations

PS — PRACTISING SKILLS DF — DEVELOPING FLUENCY PB — PROBLEM SOLVING ES — EXAM-STYLE

DF **1** **a** Draw the graphs of $y = 2x + 3$ and $y = 3 - x$ on the same pair of axes.

b Write down the co-ordinates of the point where the two lines intersect.

c Check your answer to part **b** by solving the two equations using algebra.

DF **2** **a** Draw the graphs of $2x + y = 3$ and $x - 2y = 4$ on the same pair of axes.

b Write down the co-ordinates of the point where the two lines intersect.

c Use algebra to check your answer to part **b**.

d Work out the area of the region bounded by the lines $2x + y = 3$ and $x - 2y = 4$ and the y-axis.

PB **3** Taxi companies charge a fixed amount plus an additional cost
ES per mile.

Toni's taxis	Colin's cabs
£2.50 plus £1.20 per mile.	£5.00 plus 75p per mile.

a On the same pair of axes, draw graphs to show the cost, £C, of a journey of x miles for each taxi company.

b What useful information does the point of intersection of the two graphs give you?

c Harry wants to travel 7 miles by taxi. Which company would you recommend?

DF **4** By drawing graphs, find the approximate solutions of
ES $15x + 8y = 60$
$4x - 9y = 54$

MATHEMATICS ONLY

PS **5** Line l has equation $x + y = 5$.
Line m has equation $y = 3x + 3$.
Line n has equation $y = x + 1$.
By looking at the graph, solve each
pair of simultaneous equations.

a $x + y = 5$
$y = x + 1$

b $y = 3x + 3$
$y = x + 1$

c $x + y = 5$
$y = 3x + 3$

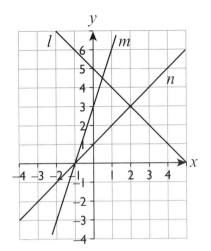

PB
ES
6 Two cars are travelling towards each other along a straight road.
The distance, d metres, from O after t seconds is given for each car by

Car A $\qquad d = 10 + 30t$ $\qquad\qquad$ Car B $\qquad d = 120 - 20t$

a On the same axes, draw graphs to show this information.

b Use your graphs to help answer these questions.

   i At what time were both cars the same distance from O?

   ii How far from O were the cars at this time?

PB
ES
7 The graph shows the speed, v metres
per second, of a car after t seconds.

a Write down the equation of this graph
in the form $v = u + at$, where u and a are
constants.

b The speed of a second car is given by the
equation $v = 80 - 2.5t$. Draw this line on a
copy of the graph.

c **i** After how many seconds are the two
cars travelling at the same speed?

   ii Estimate this speed.

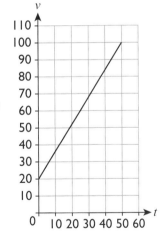

PS **8** **a** Write down the equation of
the straight line that passes
through

   i A and C $\qquad$ **ii** D and B.

b Write down the co-ordinates
of the point of intersection
of the two equations in part **a**.

c Use algebra to check your answer
to part **b**.

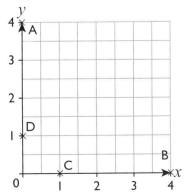

Algebra Strand 5 Working with quadratics Unit 1 Factorising quadratics

PS PRACTISING SKILLS   **DF** DEVELOPING FLUENCY   **PB** PROBLEM SOLVING   **ES** EXAM-STYLE

PS **1** Factorise each expression.

    a $x^2 + 2x$

    b $x^2 - 81$

    c $x^2 - 8x + 4x - 32$

    d $x^2 - 9x + 14$

    e $x^2 + 3x - 40$

    f $x^2 - 9$

PB **ES** **2** Amir and Winona factorised $x^2 + 5x - 6$.
Amir wrote $x^2 + 5x - 6 = (x + 2)(x + 3)$.
Winona wrote $x^2 + 5x - 6 = (x + 2)(x - 3)$.
Explain why each answer is wrong and give the correct answer.

PB **ES** **3** The area of a square is given by the expression $x^2 - 6x + 9$.
Write an expression for the side length of the square.

PB **ES** **4** The diagram shows three rectangles.

    $x + 2$ $x + 4$ $x + 7$

    $x + 1$ A $x - 1$ B x C

The area of a fourth rectangle D can be found using the equation:
Area D = Area A – Area B + Area C.
What are the dimensions of rectangle D?

DF **5** Work out each of these. Do not use a calculator.

    a $101^2 - 99^2$

    b $63^2 + 2 \times 63 \times 37 + 37^2$

    c $9^4 - 1^4$

PB **ES** **6** Bethan thinks of any number, n.

She squares her number, subtracts two times her original number from the result and then subtracts 48.

Write down and fully simplify an expression in n for her final result.

PB **ES** **7** The volume of the cuboid shown is $a^3 - 11a^2 + 30a$.

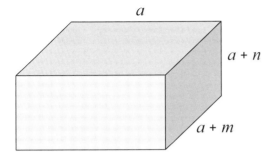

Work out the values of m and n, if $m > n$.

PB **ES** **8** Without using a calculator, work out the area of the shaded part of this shape.

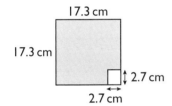

PB **ES** **9** In this cuboid:

the area of the front face is given by $p^2 + 17p + 70$

the area of the top face is given by $p^2 + 11p + 28$.

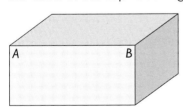

Write down an expression, in terms of p, for the length of the edge AB.

PB **ES** **10** Fully factorise this expression.

$$\frac{x^2 + 5x - 24}{x^2 - 9x - 18}$$

Algebra Strand 5 Working with quadratics Unit 2 Solving equations by factorising

PS — PRACTISING SKILLS   DF — DEVELOPING FLUENCY   PB — PROBLEM SOLVING   ES — EXAM-STYLE

PS **1** Solve these equations.

    a $(x + 1)(x + 2) = 0$

    b $(x - 4)(x - 5) = 0$

    c $x^2 + 9x = 0$

    d $x^2 - 2x - 24 = 0$

    e $x^2 = 36 - 5x$

PS **2** **a** The solutions of a quadratic equation are $x = 5$ and $x = -3$.
    Write down the quadratic equation.

    b The solutions of a quadratic equation are $y = -12$ and $y = -7$.
    Write down the quadratic equation.

PB **3** Here is Mnambi's attempt at solving $x^2 - x - 20 = 0$.

ES
$x^2 - x - 20 = 0$

$(x - 5)(x + 4) = 0$

$x = -5$ and $x = 4$

Explain the mistakes that Mnambi made and give the correct solutions.

PB **4** Rhodri thinks of a number between 1 and 10.

ES He squares the number and then subtracts his original number from the result.
His final answer is 42.
What was Rhodri's original number?

DF **5** A rectangle measures

ES $(x + 1)$ cm by $(x + 2)$ cm.
The area of the rectangle is 72 cm².
What are the dimensions of the rectangle?

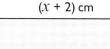
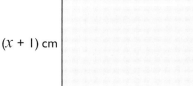

$(x + 2)$ cm

$(x + 1)$ cm

PB **ES** **6** The diagram shows a rectangle and two trapeziums drawn inside a square of side length 20 cm.

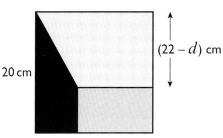

The area of the rectangle is 32 cm² and is given by $(d^2 - 4)$ cm².
Work out the area of each of the trapeziums.

PB **ES** **7** This shape is made from two identical right-angled triangles.
The total area of the shape is 135 cm².
Work out the length of the shortest side of one of these triangles.

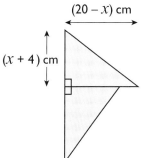

PB **ES** **8** The cuboid in the diagram has a total surface area of 246 cm².
Work out the volume of the cuboid.

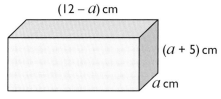

PB **ES** **9** What is the area of this right-angled triangle?

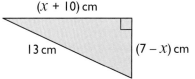

Geometry and Measures Strand 1 Units and scales Unit 7 Converting approximately between metric and imperial units

PS — PRACTISING SKILLS   **DF** — DEVELOPING FLUENCY   **PB** — PROBLEM SOLVING   **ES** — EXAM-STYLE

PB **1** Convert these volumes to litres.

   a 4 gallons

   b 9 gallons

   c 15 gallons

PB **2** Convert these distances.

   a 5 inches to cm

   b 20 miles to km

   c 1200 km to miles

DF **3** Heddwen is 163 cm tall. Linda is 5 foot 6 inches tall.
Who is taller? By how much? Give your answer in centimetres.

DF **ES** **4** A fairground ride says, 'Minimum height 4 foot 10 inches.'
Jac is 142 cm tall. Can he go on the ride? Give a reason for
your answer and show your working.

PS **5** On Monday, Tomos uses 8 gallons of fuel, costing £1.15 per litre.
How much does the fuel Tomos used on Monday cost?

PS **6** A section of a recipe says, '5 oz flour, 8 oz sugar.' Convert these to
suitable metric units.

ES **7** Lois sees a road sign, 'speed limit 30 miles per hour', which means
30 miles in one hour. How many kilometres in one hour would this be?

ES **8** A rectangle measures $2\frac{1}{2}$ inches by $1\frac{1}{4}$ inches. What are these
measurements in centimetres?

Geometry and Measures
Strand 1 Units and scales
Unit 8 Bearings

PS – PRACTISING SKILLS   DF – DEVELOPING FLUENCY   PB – PROBLEM SOLVING   ES – EXAM-STYLE

PS 1 Answer these.

   a Measure the bearing of Oxford from Bath.

   b Measure the bearing of Bath from Oxford.

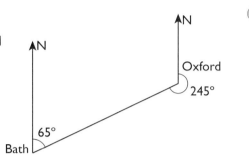

PS 2 a Write down the bearing of these compass directions.

   i West

   ii South-East.

   b Write down the compass points with these bearings.

   i 090°

   ii 225°.

DF ES 3 The diagram shows two coastguard stations, P and Q. The bearing of Q from P is 080°. There is a boat at point B. The bearing of B from P is 140°. The bearing of B from Q is 240°. Find the angle PBQ.

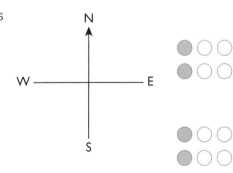

PB ES 4 An aeroplane flew 60 km from A to B, then 80 km from B to C, and then back from C to A. The bearing of B from A is 120°. The bearing of C from B is 270°.

   a Using a scale of 1 cm to represent 10 km, draw an accurate diagram to show the aeroplane's journey.

   b i What bearing must the aeroplane travel on to get from C to A?

   ii What is the distance from C to A?

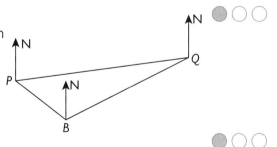

85

DF **5** Ipswich is on a bearing of 080° from Sudbury.

ES   **a** What is the bearing of Sudbury from Ipswich?

Ipswich is also North-East of Colchester.

   b Find the bearing of Ipswich from Colchester.

   c Work out the back bearing of Colchester from Ipswich.

DF **6** The diagram shows a ship, S, and its position from two lighthouses, P and Q.

Find the bearing of

   a P from Q

   b S from P

   c Q from P

   d P from S.

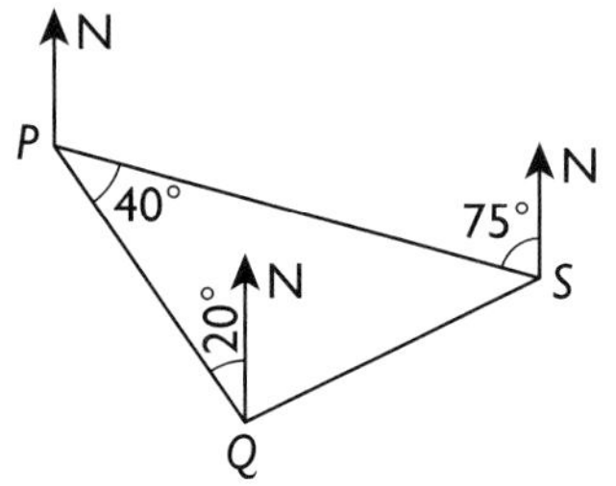

Q is South-West of S.

   e Find the bearing of S from Q.

PB **7** Ben sails his boat from Cromer on a bearing of 060° for 6 km.

ES He then changed direction and sailed on a bearing of 150° for 8 km.

   a Draw an accurate diagram to show Ben's boat trip. Use a scale of 1 cm to represent 1 km.

Ben then travels back to Cromer along a straight line.

   b What bearing does Ben travel on to get back to Cromer?

   c How far does he travel in total?

PB **8** Four buoys A, B, C and D are arranged so that

   B is on a bearing of 60° from A

ES   C is on a bearing of 105° from A

   C is on a bearing of 150° from B

   A is on a bearing of 330° from D.

Explain why ABCD is a square.

PB **9** A plane is travelling due South at 200 miles an hour. At 10 a.m.

ES the bearing of an island from the plane is 120°. At 10.30 a.m.
the bearing of the island from the plane is 060°. The plane will
not pass within 75 miles of the island. Explain why.

Geometry and Measures
Strand 1 Units and scales
Unit 9 Scale drawing

PS — PRACTISING SKILLS   DF — DEVELOPING FLUENCY   PB — PROBLEM SOLVING   ES — EXAM-STYLE

PS
ES
1 A map has a scale of 1:25 000. The distance between the church
and the garage in a village is 2 cm on the map.

   a Work out the actual distance between the church and the garage.

   The actual distance between two petrol stations is 4 km.

   b How far apart are the petrol stations on the map?

DF
2 Rosie follows these instructions when she takes part in an
orienteering competition.

     Start at the town clock and walk North for 200 metres.

     Walk East for 150 metres and then South for 100 metres.

   a Use a scale of 1 cm represents 20 m to draw a scale drawing of
Rosie's walk.

   b How far is Rosie from her starting point?

DF
ES
3 Judy wants to find the length of the sloping roof of
the end of her shed. She cannot reach the top of
the shed so draws a scale drawing of the end
of the shed.

   Find the length of the slope.

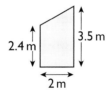

DF
4 Here is an accurate plan of a garden drawn
on a scale of 1 cm represents 4 m.

   Copy and complete the table to show
the actual measurements of the garden.

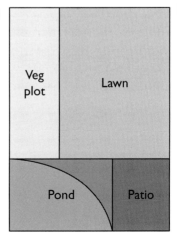

Measurement	Drawing	Actual
Length of lawn		
Width of lawn		
Radius of pond		
Width of veg plot		
Length of patio		
Width of patio		

DF **ES** **5** Tyson wants to draw a scale drawing of the town gardens.
He is going to use a sheet of paper that is 70 cm long by 44 cm wide.
The town gardens are rectangular in shape and have a length of
280 m and a width of 150 m.

Explain what scale Tyson should use to make his scale drawing
as large as possible.

PB **ES** **6** Here is a 3-D sketch of the extension Ellie wants to
add to her house. The overall height of the
extension must be 4.3 m.

Using a scale drawing, find the height of
the wall.

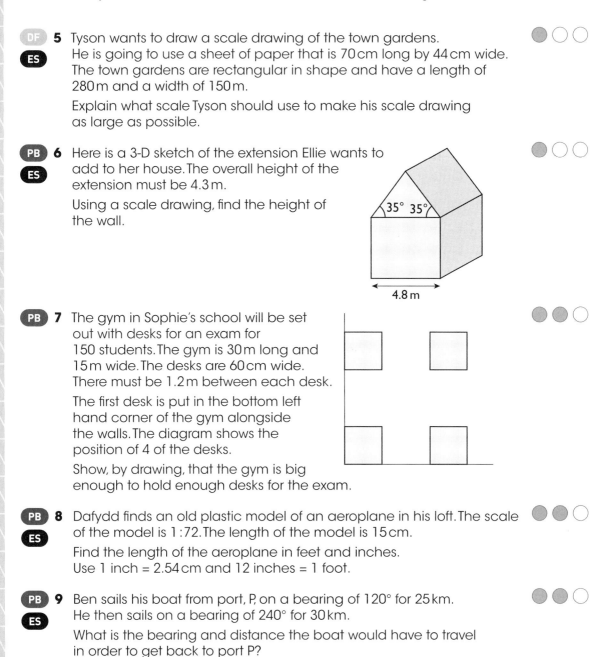

35° 35°

4.8 m

PB **7** The gym in Sophie's school will be set
out with desks for an exam for
150 students. The gym is 30 m long and
15 m wide. The desks are 60 cm wide.
There must be 1.2 m between each desk.

The first desk is put in the bottom left
hand corner of the gym alongside
the walls. The diagram shows the
position of 4 of the desks.

Show, by drawing, that the gym is big
enough to hold enough desks for the exam.

PB **ES** **8** Dafydd finds an old plastic model of an aeroplane in his loft. The scale
of the model is 1 : 72. The length of the model is 15 cm.

Find the length of the aeroplane in feet and inches.
Use 1 inch = 2.54 cm and 12 inches = 1 foot.

PB **ES** **9** Ben sails his boat from port, P, on a bearing of 120° for 25 km.
He then sails on a bearing of 240° for 30 km.

What is the bearing and distance the boat would have to travel
in order to get back to port P?

Geometry and Measures
Strand 1 Units and scales
Unit 10 Compound units

PS — **PRACTISING SKILLS**   **DF** — **DEVELOPING FLUENCY**   **PB** — **PROBLEM SOLVING**   **ES** — **EXAM-STYLE**

PS **1** Work these out.

    a Natalie drove at an average speed of 44 mph for two and a half hours. Work out how far she travelled.

    b Jason drove 330 km in 6 hours. Work out his average speed.

    c Bavna walked from her home to school at an average speed of 4.5 km/h. The school was 1.5 km from her home. Work out how long it took her.

PS **ES** **2** Mike ran 100 m in 9.8 seconds.

    a Work out his average speed in m/s.

    b Change your answer to km/h.

PS **ES** **3** Gareth's car uses 5 litres of petrol when it travels 30 miles. Susan's car uses 8 litres of petrol when it travels 50 miles.

    Whose car has the higher average rate of petrol consumption?

DF **ES** **4** Becca is going to grow a new lawn from seed. Her lawn is a rectangle 17 m long and 5 m wide. She buys a 2 kg box of grass seed which is enough to grow a 100 m² lawn.

    How many grams of grass seed will she have left?

DF **ES** **5** Toni's train leaves Bath railway station at 09:15. It arrives in London at 10:30. It travels 120 miles from Bath to London.

    Work out its average speed in mph.

DF **ES** **6** The speed of light is 186 000 miles per second. The Sun is 93 million miles from the Earth.

    Work out how long it takes a ray of light to travel from the Sun to the Earth. Give your answer in minutes and seconds.

PB **ES** **7** The speed limit on motorways in France is 130 km/h. The speed limit on motorways in the UK is 70 mph.

    Find the difference in the two speed limits. Use the fact that 5 miles = 8 km.

PB **ES** **8** Sophie has a fish pond in the shape of a cuboid. She needs to empty the full pond to clean it out.

She puts all the fish into another pond and pumps the water out at a rate of 25 litres per minute. She starts pumping out the water at 10 a.m.
At what time will the pond be empty?
Use the fact that 1 m³ = 1000 litres.

The cuboid is labelled 0.5 m, 1.5 m and 3 m.

PB **ES** **9** Mia is going to visit her mother who lives 230 miles away. She drives at an average speed of 60 mph for an hour and a half. She then stops for a 15 minute rest. She travels the remaining part of the journey at an average speed of 70 mph.

Work out Mia's overall average speed for the whole journey.

PB **10** Stuart uses this conversion graph to change between litres and gallons.

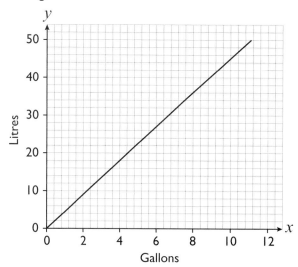

a i Change 5 gallons to litres.

ii Change 45 litres to gallons.

Stuart uses on average 0.5 litres of heating oil per hour to run his heating system. The heating system works for 14.5 hours a day for 200 days of the year. Stuart has two empty tanks in which he stores the heating oil. Tank P holds 1000 litres of oil. Tank Q holds 300 gallons of oil.

Stuart buys enough heating oil to last a whole year. He completely fills Tank P and puts the rest of the oil into Tank Q.

b How many gallons does Stuart put into Tank Q?

Stuart decides to completely fill up Tank Q as he gets a special deal on heating oil. The extra oil costs 32p per litre.

c How much does it cost to completely fill Tank Q?

Geometry and measures
Strand 1 Units and scales
Unit 11 Dimensions of formulae

PS PRACTISING SKILLS DF DEVELOPING FLUENCY PB PROBLEM SOLVING ES EXAM-STYLE

DF **1** p, q and r are all lengths.

State whether each of the following are:

lengths areas volumes none of these.

a $pq + qr$ **b** $4p + 3rq$

c $q^2 + r^2$ **d** $\pi r + 4q$

DF **2** m, n and p are all lengths.

State whether each of the following are:

lengths areas volumes none of these.

a $p^2m + m^2n$ **b** $4m - 3p^2$

c $6p^2 + \pi n^2$ **d** $\pi p^3 + 4m^2n$

DF **3** e, f and g are all lengths.

State whether each of the following are:

lengths areas volumes none of these.

a $\frac{g}{e} + f$ **b** $\frac{4e^2 + 3f^2}{g}$

c $\frac{6g^3 + \pi e^3}{f^2}$ **d** $\frac{e^3 + 4f^3}{g}$

DF **4** x, y and z are all lengths.

Write down the dimensions of each of the following.

a $\sqrt{(xy)}$ **b** $\sqrt{(x^2 - y^2)}$

c $\frac{(xy)^2}{z}$ **d** $\frac{\pi yz^2}{\sqrt{(xz)}}$

DF **5** r and h are both lengths.

One of the following expressions represents a volume.

Which one is it?

a $r^2h + \pi rh$

b $\dfrac{3rh^2}{4} - 4r^2\sqrt{h^2}$

c $4^2r^2h - 3^2rh$

d $r^3 + \dfrac{h^3}{h}$

PB **6** A shape is made from joining a hemisphere onto a cylinder.

The radius of both the hemisphere and the cylinder is r cm.

The height of the cylinder is h cm.

  a Which one of the following could be the expression for the total volume of the shape?

$\dfrac{2}{3}\pi r^2 + \pi r^2 h$ $\qquad$ $\dfrac{2}{3}\pi r^3 + \pi r^2 h$ $\qquad$ $\dfrac{2}{3}\pi r^3 + 2\pi rh$

  b Give a reason for your answer based on the dimensions of all the expressions.

PB **7** Gareth says that the surface area of the 3D shape he has made is

$rh + 2\pi h + \dfrac{1}{4}r^2$, where r and h are lengths.

  a Explain how you know Gareth's expression for what he thinks is the surface area of his 3D shape is incorrect.

  b Only one of the terms of Gareth's expression is incorrect, which term is it? And how do you know?

PB **8** Glenda says that the volume of the shape she has made from

clay is given by $\pi r^2 h + \pi rh^2 + \dfrac{(rh)^2}{rh}$

where r and h are lengths.

Could Glenda's expression possibly be correct? Give a reason for your answer.

PB **9** Harri knows that p represents a length and that r represents an area.

He is given the following formula

$G = \dfrac{1}{2}pr + p^2\sqrt{r}$

Harri needs to decide if the formula he has been given is going to calculate a length, and area or a volume.

Explain to Harri how you know what this formula could be used to calculate (a length, an area or a volume) giving detailed reasons for your answer.

Geometry Strand 1 Units and scales Unit 12 Working with compound units

PS PRACTISING SKILLS   **DF** DEVELOPING FLUENCY   **PB** PROBLEM SOLVING   **ES** EXAM-STYLE

PS
ES
1 Grass seed is sold in three sizes of box.
Which size of box is the best value for money?

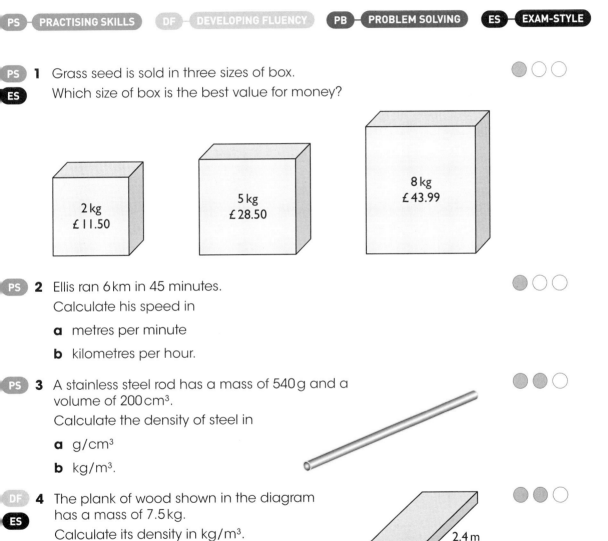

2 kg
£11.50

5 kg
£28.50

8 kg
£43.99

PS **2** Ellis ran 6 km in 45 minutes.
Calculate his speed in

  a metres per minute

  b kilometres per hour.

PS **3** A stainless steel rod has a mass of 540 g and a volume of 200 cm³.
Calculate the density of steel in

  a g/cm³

  b kg/m³.

DF
ES
4 The plank of wood shown in the diagram has a mass of 7.5 kg.
Calculate its density in kg/m³.

2.4 m

3 cm

15 cm

DF
ES
5 The graph shows Myra's car journey from her home to her mother's house.

Work out the average speed of this journey.

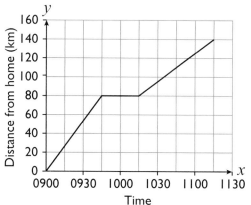

PB
ES
6 At an oil refinery, oil is stored in tanks like the one shown.

The diameter of the tank is 20 m.

The depth of the oil in the tank is 5 m.

The density of the oil is 800 kg/m³.

An oil tanker can hold up to 50 000 kg of oil.

How many of these oil tankers are needed to empty all the oil from the tank?

Show all your working.

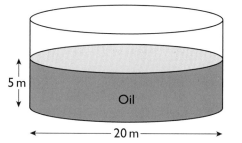

PB
7 The bronze used to make bells is an alloy of copper and tin in the ratio 4 : 1 by mass.

The density of copper is 8.96 g/cm³.

The density of tin is 7.365 g/cm³.

a A bell has a mass of 2 tonnes. Work out

  i the mass of copper

  ii the mass of tin.

b Calculate the density of the bronze.

Geometry and Measures Strand 2 Properties of shapes Unit 5 Angles in triangles and quadrilaterals

 PS PRACTISING SKILLS   DF DEVELOPING FLUENCY   PB PROBLEM SOLVING   ES EXAM-STYLE

DF **PS** **1** Find the size of angles c and d.

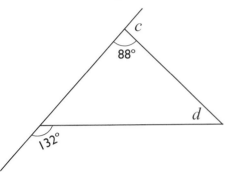

DF **PS** **2** Find the size of angle e.

ES **3** Find the size of angle a.

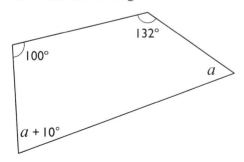

Geometry and Measures
Strand 2 Properties of shapes
Unit 6 Types of quadrilateral

PS — PRACTISING SKILLS   DF — DEVELOPING FLUENCY   PB — PROBLEM SOLVING   ES — EXAM-STYLE

PS **1** Name the quadrilaterals that have

    a two pairs of opposite sides equal and parallel

    b four equal sides

    c diagonals that cross at 90°.

PS **2** **a** Draw a trapezium that has one acute angle only.

    b Draw a kite that has one right angle only.

    c Draw an arrowhead that has one right angle only.

    d Explain why it is not possible to draw a trapezium with only one right angle.

DF **3** Draw an xy co-ordinate grid with x- and y-axes from 0 to 8.

    a Plot the points P at (2, 8) and Q at (6, 5).

        i Draw a rectangle with PQ as a diagonal.

    b Plot the point, R, at (7, 1).

        i Find the co-ordinates of S to make PQRS a parallelogram.

DF **4** Draw an xy co-ordinate grid with x-and y-axes from –2 to 8.

    a Plot the points A at (7, 2) and C at (1, 6).

        i Draw a rhombus with AC as a diagonal.

    b Plot the point B at (5, 6).

        i Find the co-ordinates of a possible position of D to make ABCD an isosceles trapezium.

DF **5** PQRS is a parallelogram.

    Find the missing angles $\hat{P}$, $\hat{Q}$ and $\hat{R}$ of the parallelogram. Explain your answer.

P   Q
65°
S   R

PB
ES
6 ABCD is an isosceles trapezium. BEC is an isosceles triangle.

AD = BC = CE

Find the size of angle D̂. Give reasons for your answer.

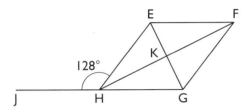

PB
ES
7 EFGH is a rhombus. JHG is a straight line.

Explain why angle EGF is 64°.

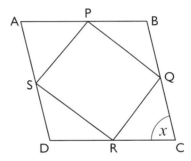

PB
ES
8 ABCD is a rhombus.

P is the midpoint of AB.
Q is the midpoint of BC.
R is the midpoint of CD.
S is the midpoint of DA.
Explain why PQRS is a rectangle.

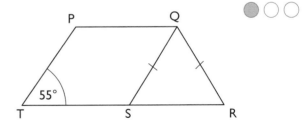

PB
ES
9 PQST is a parallelogram. QRS is an isosceles triangle.

Work out the size of angle SQR.
Give reasons for your answer.

PB
ES
10 ABCE is a parallelogram. ADE is a right-angled triangle.

Work out the size of angle EAD.
Give reasons for your answer.

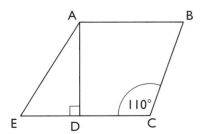

Geometry and Measures Strand 2 Properties of shapes Unit 7 Angles and parallel lines

PS – PRACTISING SKILLS   **DF** – DEVELOPING FLUENCY   **PB** – PROBLEM SOLVING   **ES** – EXAM-STYLE

PS **1** Write down the name of the marked angles.

a

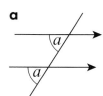

b

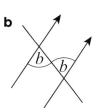

c

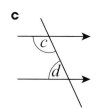

PS **2** Find the missing angles in these diagrams.
Give reasons for your answer.

a

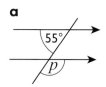

b

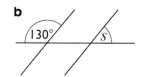

PS **3** Find the missing angles in these diagrams.

a

b

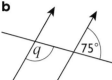

PS **4** Explain fully why the two allied (co-interior) angles marked on this diagram are supplementary.

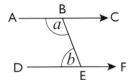

DF **5** Here is a diagram of a gate.
Find the missing angles. Give reasons for your answer.

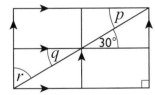

DF **ES** **6** PQRS is a rectangle. SRT is a straight line. PQTR is a parallelogram.
Find the value of angle RPS.

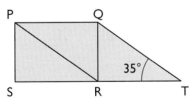

DF **ES** **7** ABDE is a parallelogram. BCD is an isosceles triangle.
Find the size of angle AED. Give reasons for your answer.

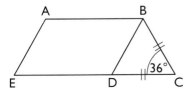

PB
ES
8 The diagram is made up from a rectangle and a parallelogram.
Find the size of the angle marked *a*.

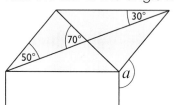

PB
ES
9 EFG is an isosceles triangle. Explain why.

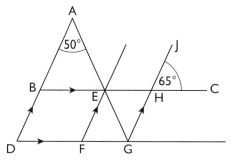

PB
ES
10 a Use this diagram to explain why the three angles of a triangle
add up to 180°. You must give reasons with your explanation.

b Write down a fact in your explanation.

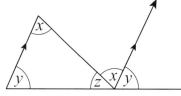

PB
ES
11 Explain why AB and CD are parallel lines. Give reasons for each
step of your explanation.

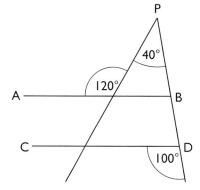

Geometry and Measures
Strand 2 Properties of shapes
Unit 8 Angles in a polygon

PS **1** Work out the size of the exterior angle and the interior angle of these regular shapes.

a
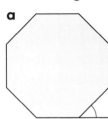

b
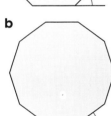

PS **2** Find the number of sides in a regular polygon that has an

  a exterior angle of 20°

  b interior angle of 108°.

DF **ES** **3** The interior angle of a regular polygon is 140°. Find the number of sides in the polygon.

PB **ES** **4** Find the size of angle x.

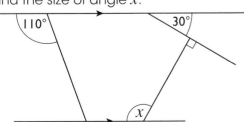

MATHEMATICS ONLY

PB **ES** **5** The diagram shows a regular octagon and a regular hexagon.
Find the size of the angle marked m.

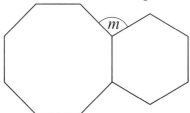

PB **ES** **6** Here is a regular dodecagon, centre O.
Explain why OPQ makes
an equilateral triangle.

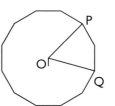

PS **ES** **7** Here is a regular octagon. Two of its
diagonals are drawn on the
shape. Find the size of angle p.

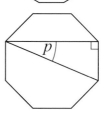

DF **ES** **8** The sum of the interior angles of a polygon is 2340°. Work out the
number of sides in the polygon.

DF **ES** **9** Three regular polygons meet at a point.
One of the polygons has an interior
angle of 60°. Another polygon has an
interior angle of 144°.

Find the number of sides in each of
the three polygons.

PB **ES** **10** The diagram shows a regular heptagon
and a regular pentagon drawn on the same
base. Work out the size of the angle marked t.

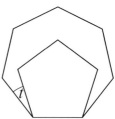

PB **ES** **11** Explain, giving reasons, why it is possible to tile a floor using regular
dodecagons and equilateral triangles that have the same length side.
You should state any assumptions that you make.

Geometry Strand 2
Properties of shapes Unit 9
Congruent triangles and proof

PS – PRACTISING SKILLS   DF – DEVELOPING FLUENCY   PB – PROBLEM SOLVING   ES – EXAM-STYLE

PS 1 State whether the triangles in each pair are congruent. ●●○
If they are, give a reason.

a
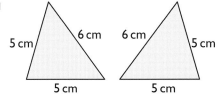

b
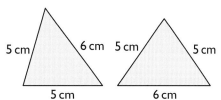

c
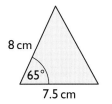

d
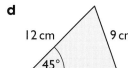

e
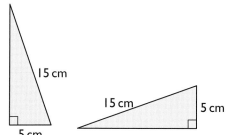

f
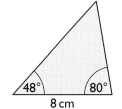
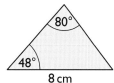

DF 2 The diagram shows a parallelogram, PQRS. ●●○
ES Prove that triangle PRS and triangle PQR are congruent.

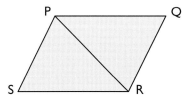

PB **ES** **3** ABCDEF is a regular hexagon.
Prove that BF = BD.

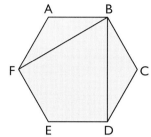

DF **ES** **4** ABCDE is a regular pentagon.
BFGC is a rectangle.
Prove that triangles ABF and
DCG are congruent.

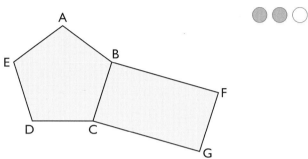

PB **ES** **5** In this diagram
AD = CD
$\angle A = \angle C = 90°$
Prove that DB bisects $\angle ABC$.

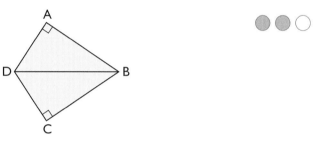

PB **ES** **6** In triangle EAD
EA = ED
$\angle AEB = \angle CED$.
Explain why AB = CD.

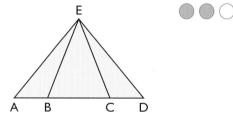

Geometry Strand 2
Properties of shapes
Unit 10 Proof using similar and congruent triangles

PS — PRACTISING SKILLS   DF — DEVELOPING FLUENCY   PB — PROBLEM SOLVING   ES — EXAM-STYLE

PS **1** For each part, state whether the two triangles are similar or congruent. ● ● ○
Give a reasons for each answer.

a

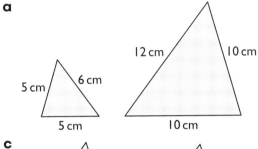

b

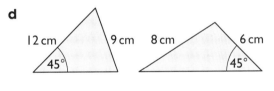

c

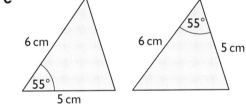

d

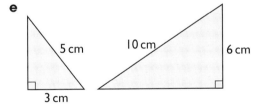

e

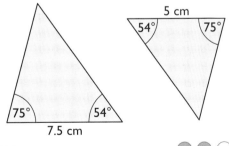

f

DF **2** Show that triangle DEF is similar to triangle GHJ. ● ● ○
ES

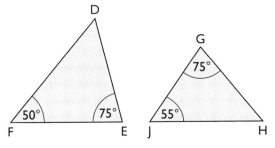

DF **3** ABD and ACE are straight lines.
BC is parallel to DE.
Prove that triangles ABC and ADE are similar.

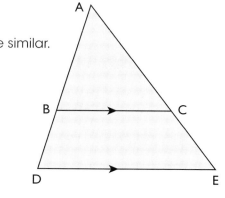

PB **4** PQRS, WXY and DEF are
ES parallel lines.
RX : XE = 2 : 3

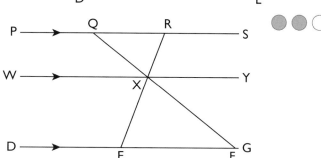

 a Prove that triangles
 QRX and EFX are similar.

 b How many times longer
 is FQ than QX?

PB **5** ABCD is a parallelogram.
ES The diagonals intersect at X.

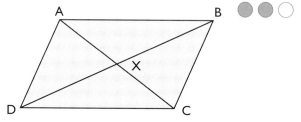

 a Prove that triangles AXD
 and BXC are congruent.

 b Show that X is the mid-point
 of AC and BD.

PB **6** PQRT and UVST are parallelograms.
PUT, TSR and TVQ are straight lines.

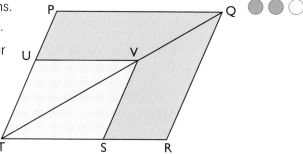

 a Prove that triangle UVT is similar
 to triangle PQT.

 b Given that PU : UT = 2 : 3, find
 the value of QV : QT.

Geometry and Measures Strand 2
Properties of shapes
Unit 11 Circle theorems

 PS — PRACTISING SKILLS   DF — DEVELOPING FLUENCY   PB — PROBLEM SOLVING   ES — EXAM-STYLE

PS   **1**   Find the size of the angle marked with a letter in each diagram.
Give a reason for each of your answers.

a
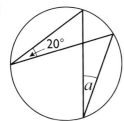

b
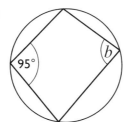

c
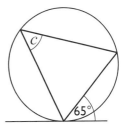

PS   **2**   Each circle in this question has a centre O. Find the size of the angle marked with a letter. Give a reason for each of your answers.

a
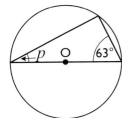

b
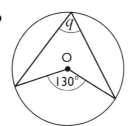

c
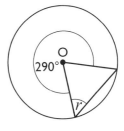

DF   **3**   The diagram shows a circle centre O. PA and PB are tangents to the
ES   circle.

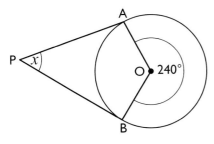

Find the size of the angle marked x.

PB **ES** **4** P, Q, R and S are points on the circumference of a circle centre O. Angle PRQ = 50°.

Find the size of the angle marked y. Give reasons for each stage of your working.

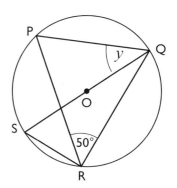

DF **ES** **5** D, E and F are points on the circumference of a circle centre O. Angle DOF = 140°.

Work out the size of the angle marked g. Give reasons for each stage of your working.

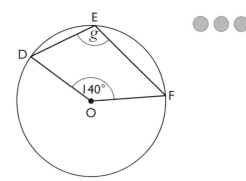

PB **ES** **6** TP and TR are tangents to the circle centre O.

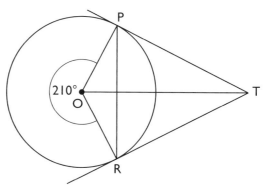

Find the size of angle PTR. Give reasons for each stage of your working.

PB **ES** **7** A, B and C are points on the circle centre O. Angle ABC is $x°$. TA and TC are tangents to the circle.

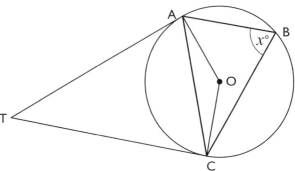

Prove that TA = TC.

MATHEMATICS ONLY

PB **8** B, C, D and E are points on the circumference of the circle centre O.
ES EB is parallel to DC. XD is a tangent to the circle at D.

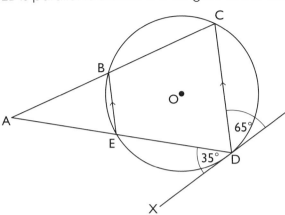

Using the information on the diagram prove that triangle ABE is isosceles.

PB **9** AB, BC and CA are tangents to the circle at P, Q and R respectively.
ES Angle B = $2x°$. Angle C = $2y°$.

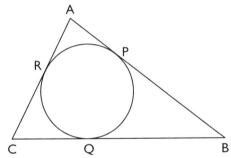

Find an expression in terms of x and y for the size of angle PQR.

PB **10** The circle centre O has a radius of 7 cm. The circle centre P has a
ES radius of 10 cm. AB is a tangent to both circles.

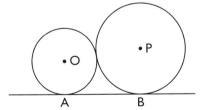

a Find the length of AB.

b What assumptions have you made in your solution?

Geometry and Measures
Strand 3 Measuring shapes
Unit 4 Area of circles

PS — PRACTISING SKILLS   **DF** — DEVELOPING FLUENCY   **PB** — PROBLEM SOLVING   **ES** — EXAM-STYLE

PS **1** Find the area of these circles with radius

    a 5 cm

    b 7 m

    c 3.2 cm.

    Give your answers correct to one decimal place.

> Don't forget
> Area $= \pi \times r^2$

PS **2** Find the area of these circles with diameter

    a 6 cm

    b 5 m

    c 0.8 km.

    Give your answers correct to one decimal place.

PS **3** **a** A circle has an area of 15.7 cm². Find the radius of the circle.
    Give your answer correct to two decimal places.

    b A circle has an area of 1 m². Find the diameter of the circle.
    Give your answer to the nearest centimetre.

DF **4** **a** A circle has an area of 50 cm². Find the circumference of the circle.
    Give your answer correct to one decimal place.

    b A circle has a circumference of 314 cm. Find the area of the circle.
    Give your answer correct to the nearest metre².

DF **ES** **5** Mo makes a semicircular flower bed. The radius
of the flower bed is 1.5 m.
Find the area of the flower bed. Give your answer
correct to 2 decimal places.

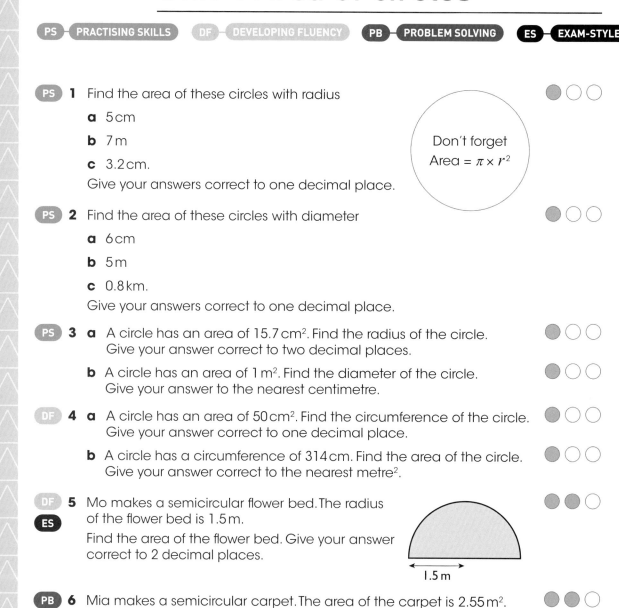

1.5 m

PB **ES** **6** Mia makes a semicircular carpet. The area of the carpet is 2.55 m².
Find the perimeter of the rug. Give your answer to the nearest cm.

PB
ES
7 Naga feeds the fish in her fishpond. The fishpond is in the shape of a quarter circle. For every square metre of the area of the pond she gives the fish 25 g of fish food each day.

How many grams of fish food does she need to give the fish? Give your answer to the nearest gram.

2.5 m

PB
ES
8 The diagram shows a circular pond surrounded by a path. The pond has a radius of 2.5 m. The path is 1 metre wide. The path is going to be covered with wood bark. Wood bark is sold in bags that cost £2.99 each. Each bag contains enough bark to cover 0.75 m².

Work out the cost of the wood bark needed to cover the path.

PB
ES
9 Percy is making a lawn. The lawn is in the shape of half a ring made from two semicircles.
The semicircles have the same centre.
The radius of the large semicircle is 15 m.
The radius of the small semicircle is 5 m.

Percy buys 1 kg boxes of lawn seed at £9 per box. Each kg of lawn seed covers 25 m².

How much does it cost Percy to buy the grass seed?

15 m   5 m

DF
ES
10 Paul makes a brooch. He makes it from a semicircle and two quarter circles of metal. The diameter of the semicircle is 4.5 cm. The radius of the large quarter circle is 2.5 cm. The radius of the small quarter circle is 1.5 cm.

Find the area of the brooch.

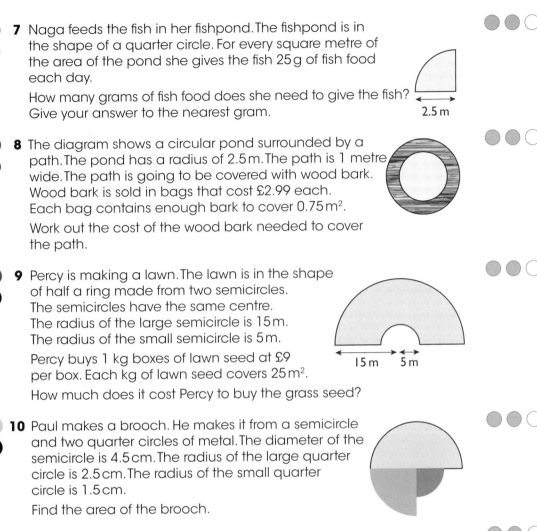

PB
ES
11 This shape is made from a right-angled triangle, a semicircle and a quarter circle.

Find the area of this shape.

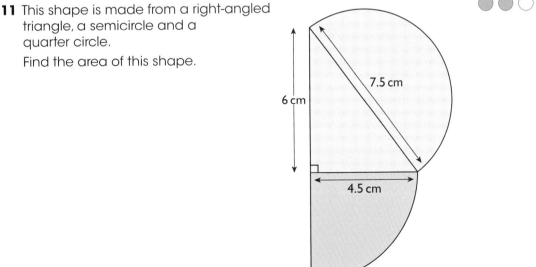

6 cm

7.5 cm

4.5 cm

Geometry Strand 3
Measuring shapes Unit 5
Pythagoras' theorem

PS — PRACTISING SKILLS   DF — DEVELOPING FLUENCY   PB — PROBLEM SOLVING   ES — EXAM-STYLE

PS 1 Work out the length of the hypotenuse in each triangle.
Give your answers correct to 1 decimal place.

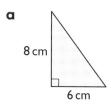

a 8 cm 6 cm

b 6 cm 4 cm

c 3 cm 4.5 cm

PS 2 Work out the length of the unknown side in each triangle.
Give your answer correct to 2 decimal places.

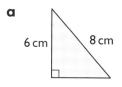

a 6 cm 8 cm

b 6 cm 4 cm

c 3 cm 4.5 cm

DF 3 Without using a calculator, work out the length of the unknown side
in each triangle.

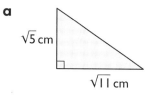

a √5 cm √11 cm

b √35 cm √10 cm

c √44 cm 12 cm

DF 4 Which of the three triangles are right-angled triangles?
Explain your answer.

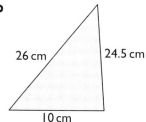

a 2 cm 2.5 cm 1.5 cm

b 26 cm 24.5 cm 10 cm

c 4.5 cm 4.5 cm 6.4 cm

DF **ES** **5** ABCD is a rectangle.

Work out the length of the diagonal BD.

Give your answer correct to 3 significant figures.

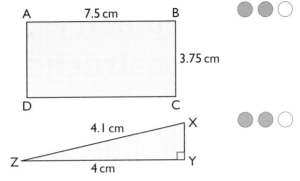

DF **6** For triangle XYZ, work out

 a the perimeter

 b the area.

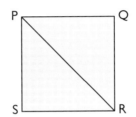

DF **ES** **7** PQRS is a square. The diagonal is 16 cm.

Work out the perimeter of the square.

Give your answer correct to 3 significant figures.

PB **8** A square is drawn with its vertices on the circumference of a circle. The diagonal of the square is 8 cm.

Work out the area of the shaded part of the diagram, giving your answer correct to 3 significant figures.

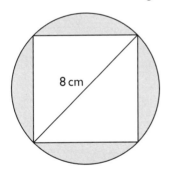

PB **ES** **9** Calculate the area of the field shown in this diagram.

Give your answer in hectares to 3 significant figures.

(1 hectare = 10000 m²)

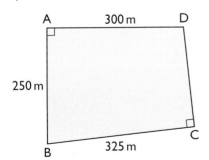

Geometry Strand 4
Construction Unit 3
Constructions with a pair of compasses

PS — PRACTISING SKILLS   DF — DEVELOPING FLUENCY   PB — PROBLEM SOLVING   ES — EXAM-STYLE

PS **1** Use a ruler and a pair of compasses to construct each triangle accurately.

a
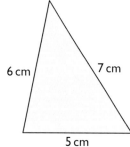
6 cm   7 cm
5 cm

b
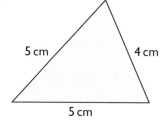
5 cm   4 cm
5 cm

c
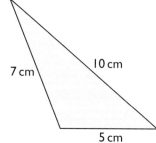
7 cm   10 cm
5 cm

PS **2** Use a ruler and a pair of compasses to construct each triangle accurately.

a
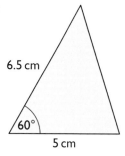
6.5 cm
60°
5 cm

b
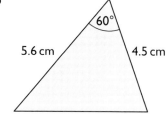
60°
5.6 cm   4.5 cm

c
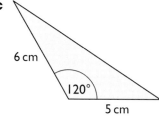
6 cm
120°
5 cm

PS **3** Use a ruler and a pair of compasses to construct each angle.

a 90°

b 120°

c 45°

d 30°

DF **4 a** Use a ruler and a pair of compasses to construct triangle ABC.

    b Construct the perpendicular bisector of each of the three sides.

    c Mark the point M where the three perpendicular bisectors meet.

    d Draw a circle with centre M that passes through the points A, B, and C.
       This is called the circumscribed circle of the triangle.

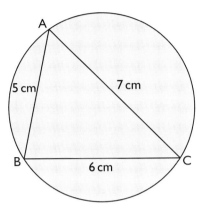

DF **5 a** Use a ruler and a pair of compasses to construct triangle DEF.

    b Bisect the exterior angles BDE and GED.

    c Mark the point M where the angle bisectors meet.

    d Draw a circle with centre M to touch BD, DE and EG.
       This is called an escribed circle of the triangle.

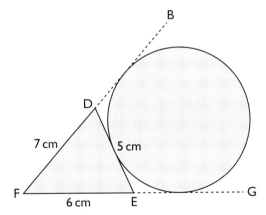

PB **ES** **6 a** Use a ruler and a pair of compasses to construct a triangle PQR with
       PQ = 8 cm, QR = 6.8 cm and angle PQR = 45°.

    b Mark the point A on PQ so that PA = AQ.

    c Construct a rectangle ABCD so that B lies on PR and both C and D
       lie on QR.

Geometry Strand 4
Construction Unit 4 Loci

PS PRACTISING SKILLS   **DF** DEVELOPING FLUENCY   **PB** PROBLEM SOLVING   **ES** EXAM-STYLE

PS **1** Draw a co-ordinate grid on 2mm graph paper.

Draw the x-axis from –6 to 6.

Draw the y-axis from –5 to 7.

Draw the locus of the points that are

a 4cm from (1, 1)

b 2cm from the line joining (–3, 3) and (–3, –2)

c the same distance from the points (1, 5) and (5, 1)

d the same distance from the lines joining (–4, 4) to (0, –4) and (0, 4) to (–4, –4).

PS **2** The diagram shows Fflur's garden.

Fflur wants to plant a new tree in her garden.

The tree will be planted:

- nearer to RQ than RS

- less than 8m from Q.

a Draw the diagram accurately, using a scale of 1cm = 2m.

b Shade the region where Fflur could plant her new tree.

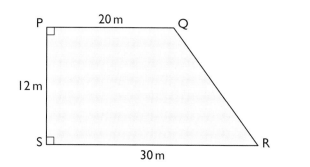

PB **3** The diagram shows the positions of Colchester and Ipswich.

Ipswich is 18 miles north east of Colchester.

A company want to build a new hotel such that

- it is nearer Ipswich than Colchester

- it is less than 12 miles from Colchester.

a Draw the diagram accurately, using a scale of 1cm = 2 miles.

b Shade the region where the hotel could be built.

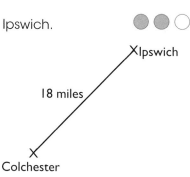

PB **4** The diagram shows a car park that measures 75m by 55m.

Cars must not be parked within 20m of W or within 15m of XY.

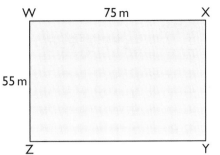

a Draw the car park accurately, using a scale of 1cm = 10m.

b Shade the region where the cars should not be parked.

DF
ES
5 PQRS is a square piece of card placed on a straight line.

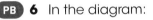

The card is first rotated 90° clockwise about R.

It is then rotated 90° clockwise about Q.

Finally it is rotated 90° clockwise about P.

Draw the locus of the vertex S.

PB
ES
6 In the diagram:

- P and Q are buoys.
  P is 750m due North of Q
  and Q is 1km NE of H.

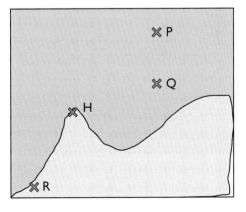

- H is 1km on a bearing of 020° from R.

  Ewan has to steer his boat along a course from port R between the buoys at P and Q. He must stay at least 300m away from H.

  He wants to sail due North from R and then onto a course on the perpendicular bisector of P and Q.

a Draw an accurate diagram of Ewan's boat trip, using a scale of 1cm = 100m.

b Will he pass too close to H?
Explain your answer.

Geometry and Measures
Strand 5 Transformations
Unit 3 Translation

PS — **PRACTISING SKILLS** DF — **DEVELOPING FLUENCY** PB — **PROBLEM SOLVING** ES — **EXAM-STYLE**

PS
DF
PB
ES

1 Copy the diagram and answer the questions.

    a Translate the triangle $\begin{pmatrix} 2 \\ -3 \end{pmatrix}$.

    b Write down the co-ordinates of each of the vertices of your translated triangle.

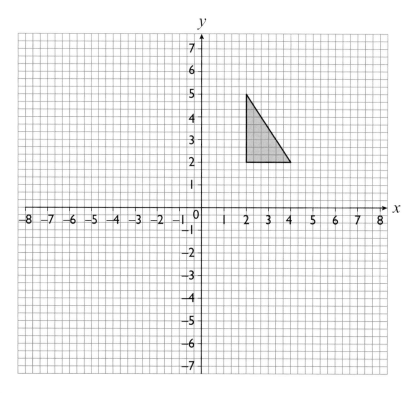

PS
DF
PB
ES

2 Copy the diagram and answer the questions.

    a Translate the triangle $\begin{pmatrix} -1 \\ -3 \end{pmatrix}$.

    b Write down the co-ordinates of each of the vertices of your translated triangle.

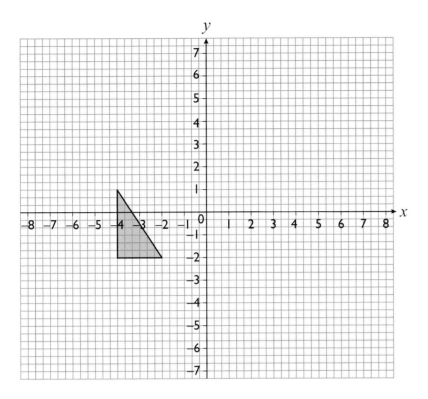

PS
DF
PB
ES

3 Copy the diagram and answer the questions.   ● ○ ○

a Translate the triangle $\binom{2}{8}$.

b Write down the co-ordinates of each of the vertices of your translated triangle.

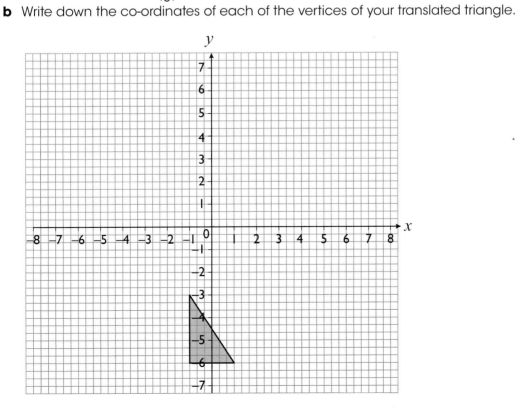

Geometry and Measures
Strand 5 Transformations
Unit 4 Reflection

PS — PRACTISING SKILLS   DF — DEVELOPING FLUENCY   PB — PROBLEM SOLVING   ES — EXAM-STYLE

DF **1** Copy the diagram and answer the questions.   ● ○ ○

PB **a** Reflect the triangle shown in the x-axis.

b Write down the co-ordinates of each of the vertices of your reflected triangle.

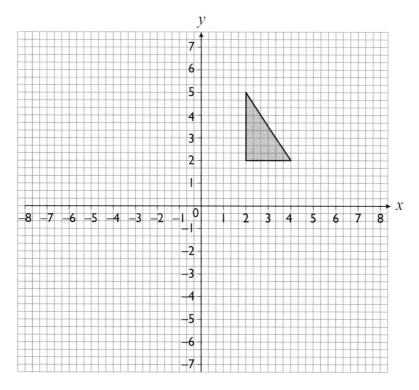

DF **2** Copy the diagram and answer the questions.

PB

    a Reflect the triangle shown in the y-axis.

    b Write down the co-ordinates of each of the vertices of your reflected triangle.

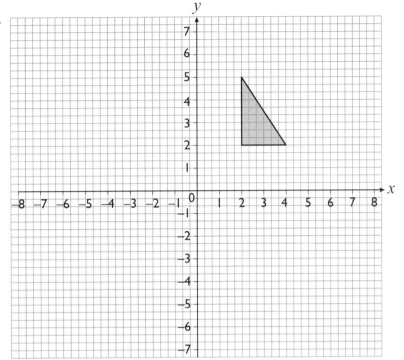

PS **3** Copy the diagram and answer the questions.

ES

    a Reflect the triangle shown in the line $x = -1$.

    b Write down the co-ordinates of each of the vertices of your reflected triangle.

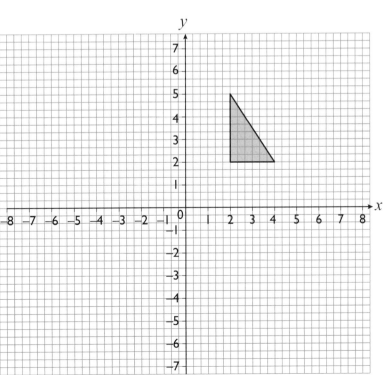

PS
ES

4 Copy the diagram and answer the questions.

   a Reflect the triangle shown in the line $y = x$.

   b Write down the co-ordinates of each of the vertices of your reflected triangle.

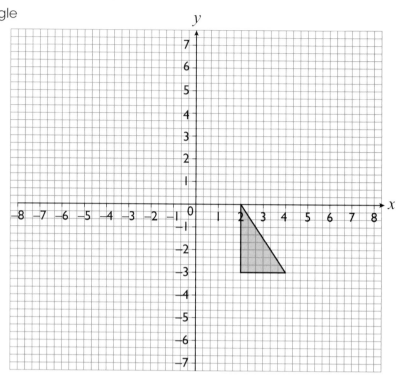

PS
ES

5 Copy the diagram and answer the questions.

   a Reflect the triangle shown in the line $y = -x$.

   b Write down the co-ordinates of each of the vertices of your reflected triangle.

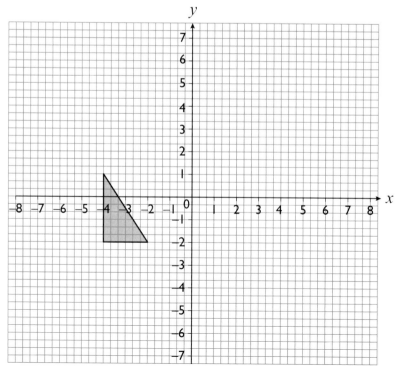

Geometry and Measures
Strand 5 Transformations
Unit 5 Rotation

 PS — PRACTISING SKILLS   DF — DEVELOPING FLUENCY   PB — PROBLEM SOLVING   ES — EXAM-STYLE

DF **PB** **1** Copy the diagram and answer the questions.

    a Rotate the triangle shown 180° about the origin (0, 0).

    b Write down the co-ordinates of each of the vertices of your rotated triangle.

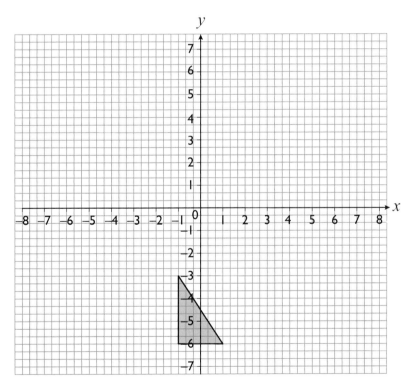

DF **PB** **2** Copy the diagram and answer the questions.

    a Rotate the triangle shown 90° clockwise about the origin (0, 0).

    b Write down the co-ordinates of each of the vertices of your rotated triangle.

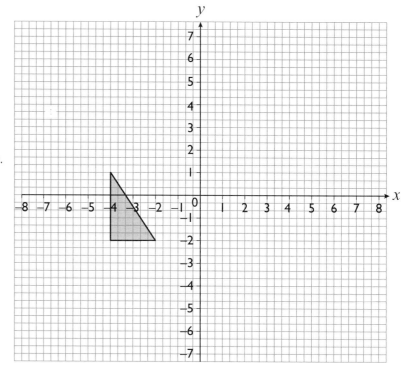

DF **PB** **3** Copy the diagram and answer the questions.

    a Rotate the triangle shown 90° anticlockwise about the origin (0, 0).

    b Write down the co-ordinates of each of the vertices of your rotated triangle.

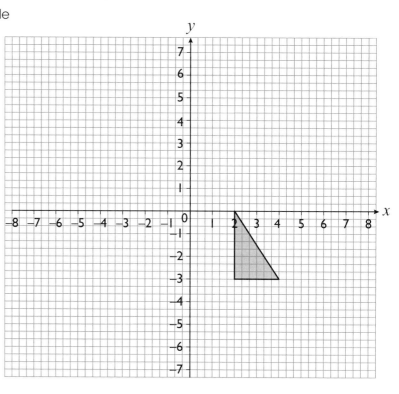

PS
ES

4 Copy the diagram and answer the questions.

● ○ ○

a Rotate the triangle shown 180° about the point (1, 2).

b Write down the co-ordinates of each of the vertices of your rotated triangle.

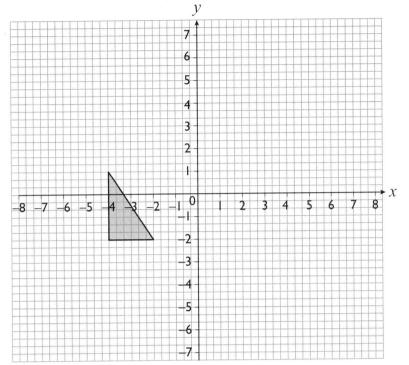

PS
ES

5 Copy the diagram and answer the questions.

● ○ ○

a Rotate the triangle shown 90° anticlockwise about the point (−1, 2).

b Write down the co-ordinates of each of the vertices of your rotated triangle.

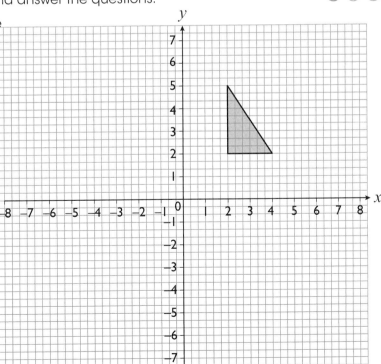

125

PS 6 Copy the diagram and answer the questions.

ES

a Rotate the triangle shown 90° anticlockwise about the point (–3, –2).

b Write down the co-ordinates of each of the vertices of your rotated triangle.

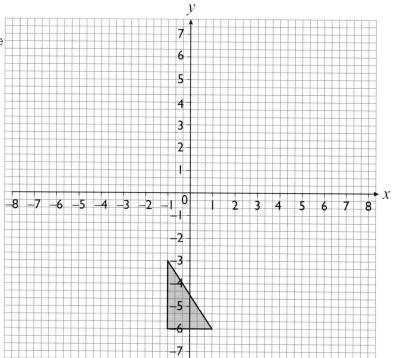

PS 7 Copy the diagram and answer the questions.

ES

a Rotate the triangle shown 90° anticlockwise about the point (–2, 1).

b Write down the co-ordinates of each of the vertices of your rotated triangle.

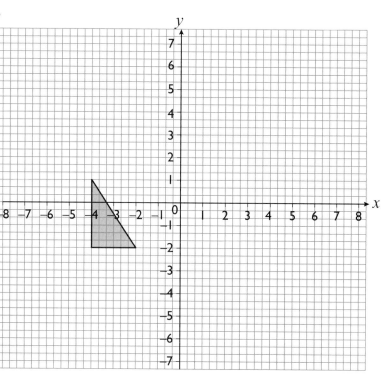

Geometry and Measures
Strand 5 Transformations
Unit 6 Enlargement

PS — PRACTISING SKILLS   DF — DEVELOPING FLUENCY   PB — PROBLEM SOLVING   ES — EXAM-STYLE

PS **1** Here is a grid of squares.

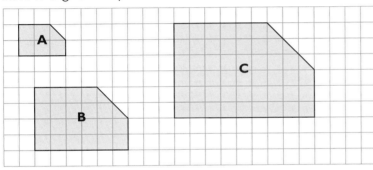

Write down the scale factor of the enlargement that takes

a A to B

b A to C

c B to C.

DF **2** Enlarge triangle P by
PB

a scale factor 2 centre (−1, 0)

b scale factor 3 centre (−4, 2).

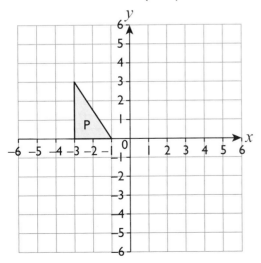

DF
PB

3 Make three copies of this diagram. Enlarge triangle T by

   a scale factor 3 centre (4, 3)

   b scale factor 2 centre (2, 2)

   c scale factor 1.5 centre (3, 1).

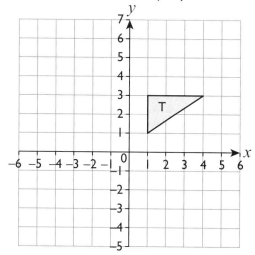

DF **4** Describe fully the single transformation that maps

   a triangle T onto triangle Q

   b triangle T onto triangle R.

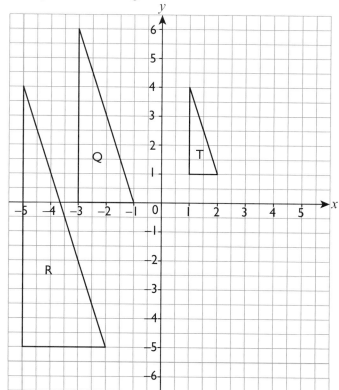

PB
ES
5 Here is a small photograph. The small photograph is enlarged by scale factor 3 to make a large photograph.

What is the perimeter of the large photograph?

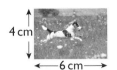

4 cm

←—6 cm—→

PB
ES
6 Here is a photograph. Jill makes enlargements of the photograph.

6 cm

←——— 8 cm ———→

Enlargement A

←——————— 32 cm ———————→

Enlargement B

←——————— 16 cm ———————→

a Find the scale factor of the enlargement of Enlargement A.

b Find the width of Enlargement A.

c Work out the perimeter of Enlargement B.

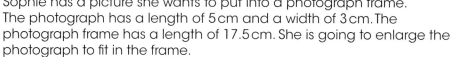

PB **ES** **7** Sophie has a picture she wants to put into a photograph frame. The photograph has a length of 5 cm and a width of 3 cm. The photograph frame has a length of 17.5 cm. She is going to enlarge the photograph to fit in the frame.

What width does the frame have to be to fit in the enlarged photograph?

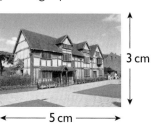

3 cm

5 cm

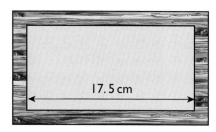

17.5 cm

DF **8** Enlarge triangle P by

   a scale factor $\frac{1}{2}$ centre $(-4, 4)$

   b scale factor $\frac{1}{3}$ centre $(-4, -5)$.

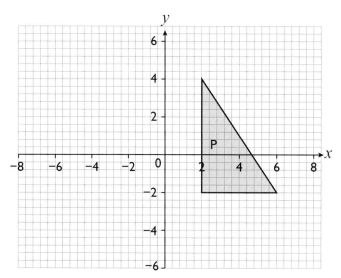

Geometry Strand 5
Transformations Unit 7
Similarity

PS **1** Work out the lettered lengths in each pair of similar triangles. ● ● ○

a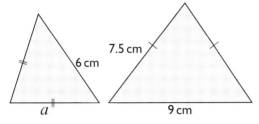

b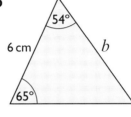

c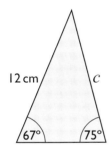

d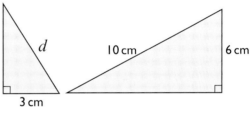

PS **2** XYZ and PQR are two similar triangles. Work out the length of ● ● ○

a XY

b PR.

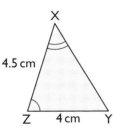

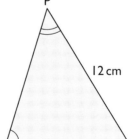

DF **3** Work out the lettered lengths in each pair of similar shapes. ● ● ○

a

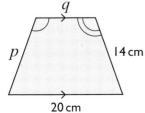

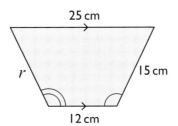

b

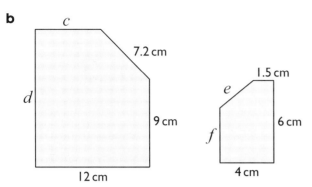

DF **ES** **4** ABC and ADE are two similar triangles.

BC is parallel to DE.

Work out the length of

a BC

b AE.

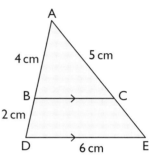

DF **ES** **5** Triangles DEF and HGJ are similar.

Work out the length of

a DE

b JH.

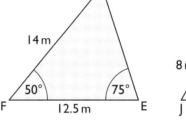
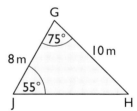

DF **ES** **6** In this diagram, PQ is parallel to RS.

Work out the length of

a RT

b PT.

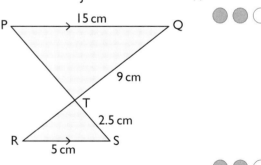

PB **ES** **7** The diagram shows a set of supports of two sizes made to support shelves.

Work out the value of

a x

b y.

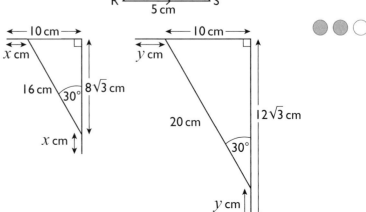

Geometry Strand 5
Transformations Unit 8
Trigonometry

PS — **PRACTISING SKILLS**   DF — DEVELOPING FLUENCY   PB — **PROBLEM SOLVING**   ES — **EXAM-STYLE**

PS **1** Work out the length of the lettered side in each right-angled triangle.
Give your answers correct to 1 decimal place.

a

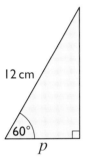

12 cm

60°

p

b

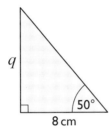

q

50°

8 cm

c

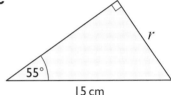

r

55°

15 cm

PS **2** For each triangle, work out the value of θ.
Give your answer correct to 1 decimal place.

a

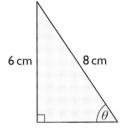

6 cm   8 cm

θ

b

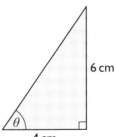

6 cm

θ

4 cm

c

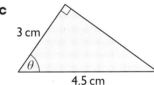

3 cm

θ

4.5 cm

PS **3** Work out the length of the lettered side in each right-angled triangle.
Give your answers correct to 2 decimal places.

a

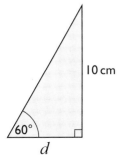

10 cm

60°

d

b

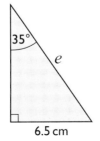

35°

e

6.5 cm

c

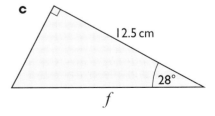

12.5 cm

28°

f

133

DF **ES** **4** Work out the perpendicular height h of this triangle.
Give your answer correct to 3 significant figures.

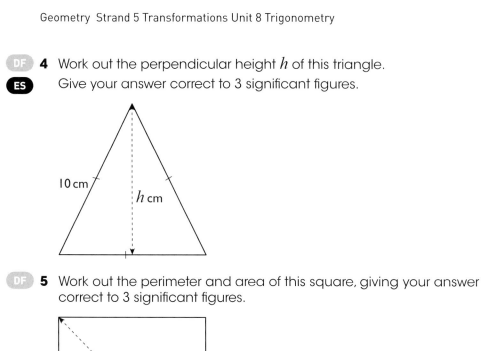

DF **5** Work out the perimeter and area of this square, giving your answer correct to 3 significant figures.

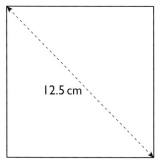

12.5 cm

DF **ES** **6** Betsan is 80 m from a TV mast on horizontal ground.

She measures the angle of elevation of the top of the TV mast as 40°.

Work out the height of the TV mast, giving your answer correct to 1 decimal place.

DF **ES** **7** Osian is standing 30 m from the tree and his angle measurer is 120 cm above the ground.

He measures the angle of elevation to the top of tree as 35°.

Work out the height of the tree, correct to 3 significant figures.

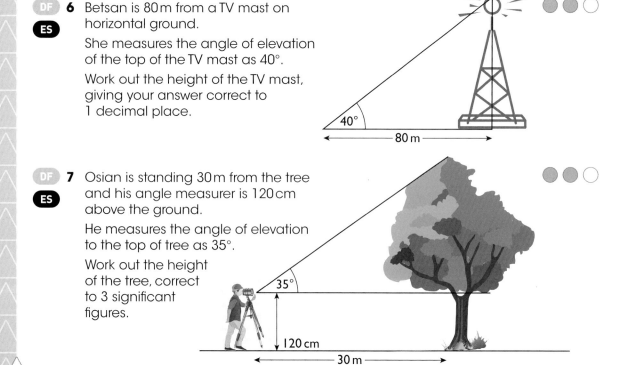

PB **ES** **8** The diagram shows a framework made from five rods.

The rectangle has a length of 12 m.

The diagonal makes an angle of 25° with the base of the rectangle.

Work out the total length of the five rods in the framework.

Give your answer correct to 3 significant figures.

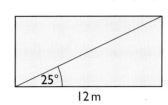

12 m

PB **ES** **9** The longest diagonal of a rhombus is 10 cm.

This diagonal makes an angle of 30° with the base of the rhombus.

Work out the perimeter of the rhombus, giving your answer correct to 3 significant figures.

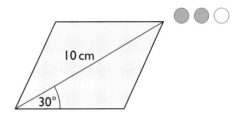

PB **ES** **10** Alfie takes his boat to check two offshore windmills at W and M.

He leaves the harbour H and travels due East for 12 km to W and then 5 km due North to M.

On what bearing must he travel to get directly back to H?

Give your answer to the nearest degree.

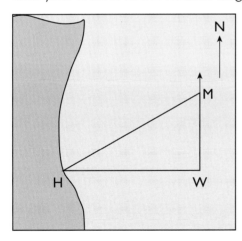

PB **ES** **11** Kirsty wants to find the width of a river.

She stands at the top of the tower, T.
The base of the tower is at A.

ABC is a straight line.

Angle TAB is a right angle.

The angle of depression of:

· B from T is 40°

· C from T is 30°.

The base of the tower is 100 m from B.

Work out the width of the river.

Give your answer correct to 3 significant figures.

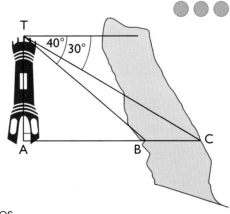

Geometry and Measures Strand 6 Three-dimensional shapes Unit 3 Volume and surface area of cuboids

PS — PRACTISING SKILLS   DF — DEVELOPING FLUENCY   PB — PROBLEM SOLVING   ES — EXAM-STYLE

PS **1** Work out the volume of these cuboids.

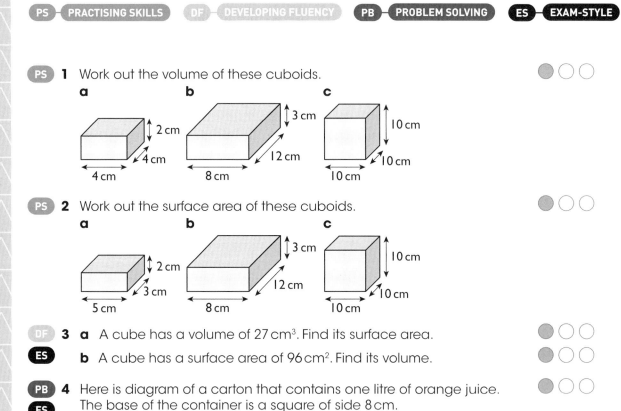

a
2 cm
4 cm
4 cm

b
3 cm
12 cm
8 cm

c
10 cm
10 cm
10 cm

PS **2** Work out the surface area of these cuboids.

a
2 cm
3 cm
5 cm

b
3 cm
12 cm
8 cm

c
10 cm
10 cm
10 cm

DF **3 a** A cube has a volume of 27 cm³. Find its surface area.

ES **b** A cube has a surface area of 96 cm². Find its volume.

PB **4** Here is diagram of a carton that contains one litre of orange juice.

ES The base of the container is a square of side 8 cm.

Find the least height of the container.

Orange Juice
I litre

1 litre = 1000 cm³

5 Phil has a box of toy bricks. Each toy brick is a cube with side 5 cm.
The box is full of toy bricks.
What is the greatest number of toy bricks that can fit into the box?

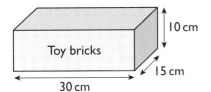

6 Tea is sold in boxes. The boxes are 12 cm tall. The base is a square
of side 5 cm. The boxes of tea are delivered to shops in cartons. The
cartons are cuboids with length 60 cm, width 30 cm and height 36 cm.
Work out the greatest number of boxes of tea that will fit into a carton.

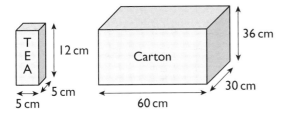

7 An oil tank is in the shape of a cuboid.
The oil tank is 2.5 m by 1.5 m by 60 cm.

a Find the volume of the oil tank.

The oil tank is half full of oil.

b How much oil needs to be added to
the tank to fill it completely?

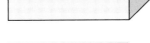

$1\,\text{m}^3 = 1000$ litres

8 Nick is going to paint his swimming pool. The pool stands on flat
ground and is in the shape of a cuboid. He is going to paint the four
outside faces and all the inside faces.

The pool has dimensions 15 m by 8 m by 1.2 m. The paint Nick is going
to use covers 15 m² with 1 litre of paint.
How many litres of paint will Nick need to buy?

9 Jill designs a box with a lid. The box is to hold wooden discs which
have a radius of 6 cm and a height of 0.5 cm. Jill's box has to hold
up to 120 discs.
Work out the dimensions of a box that Jill could use.

10 Penny has a shed with a flat roof. The roof is a rectangle with length
1.8 m and width 1.2 m. One night 2.5 cm of rain fell on the roof and
was collected in a container that is in the shape of a cuboid.

The container has a height of 1.5 m and has a square base of
side 30 cm. At the beginning of the night there was 10 cm of rain
in the container.
Work out the height of the water in the container at the end of
the night.

137

Geometry and Measures Strand 6 Three-dimensional shapes Unit 4 2-D representations of 3-D shapes

PS — PRACTISING SKILLS   DF — DEVELOPING FLUENCY   PB — PROBLEM SOLVING   ES — EXAM-STYLE

PB **1** Here is a 3-D drawing of Graham's garage. The front face and back face are rectangles. The other two vertical faces are trapeziums.

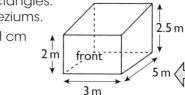

On squared paper, using a scale of 1 cm to represent 1 m

a draw a plan of the garage

b draw the front elevation of the garage

c draw the side elevation of the garage from the direction shown by the arrow.

DF **2** Here are some mathematical shapes. They have all got square bases of side 5 cm. On squared paper sketch for each shape

a the plan

b the front elevation

c the side elevation.

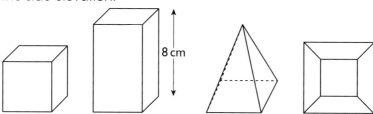

PS **3** Here are the plan, front elevation and side elevation of a 3-D shape.

ES Draw a sketch of the 3-D shape.

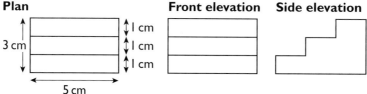

DF
ES

4 Here are the plan, front elevation and side elevation of a shape that has been made from centimetre cubes.
Draw a sketch of the 3-D solid made from these cubes.
Use isometric paper.

Plan

Front elevation

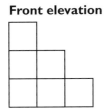

Side elevation

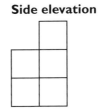

DF

5 Here is a shape made from a cuboid and a triangular prism. The cuboid has dimensions 6 cm by 4 cm by 4 cm. The triangular prism has a square base and has a height of 6 cm.

a Draw the plan and front elevation for the 3-D shape.

b Draw the side elevations from the direction of
  i the black arrow
  ii the white arrow.

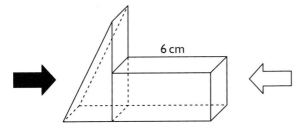

6 cm

DF
ES

6 Here are the plan, front and side elevation of a 3-D shape.
Draw a sketch of the 3-D shape.

Plan

Side elevation

Front elevation

PB
ES

7 Here is a drawing of Sid's shed.
The shed is made from 4 vertical walls and a rectangular sloping roof.

a Draw the side elevation of the shed.

b Find the area of the roof of the shed.

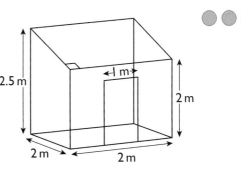

2.5 m

1 m

2 m

2 m

2 m

Geometry and Measures Strand 6 Three-dimensional shapes Unit 5 Prisms

PS — PRACTISING SKILLS   DF — DEVELOPING FLUENCY   PB — PROBLEM SOLVING

ES — EXAM-STYLE

PS **1** Here are some prisms. Find their volume.

a   **b**   **c**

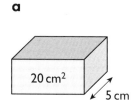

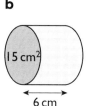

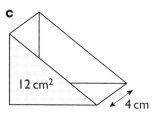

PS **2** Here are some prisms. Find their total surface area.

a   **b**   **c**

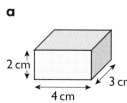

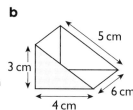

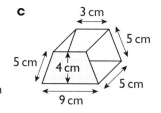

PS **3** Here is a cylinder.

ES

    a Find the volume of the cylinder.

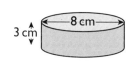

    b Find the total surface area of the cylinder.

DF **4** Answer these.

    a A hexagonal prism has a cross-sectional area of $35\,cm^2$ and a length of $5\,cm$. Find the volume of the hexagonal prism.

    b A cylinder has a volume 1 litre. The area of the circular base is $50\,cm^2$. Find the height of the cylinder.

    c An octagonal prism has a volume of $24\,cm^3$. The length of the prism is $3\,cm$. Find the area of the base of the prism.

PB
ES
5 Here is a sketch of Fred's shed. It is in the shape of a pentagonal prism. Fred is going to paint the 6 outside faces of the shed. 1 litre of the paint he uses covers 15 m².

How many litres of paint does he need?

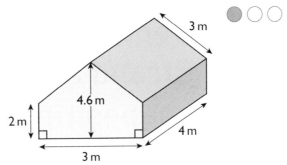

DF
6 Siân makes a plant pot holder for her garden from a cylindrical block of wood. The cylinder has a diameter of 30 cm and a height of 30 cm. Siân cuts a hole with radius 10 cm to a depth of 20 cm into the cylinder.

What volume of wood is left from the original cylinder?

PB
ES
7 Susi is going to grow vegetables in a container in her garden. The container is in the shape of a prism. The ends of the prism are trapeziums.

Susi plans to fill the container completely with compost to a depth of 60 cm. Compost is sold in 65 litre bags.

How many bags of compost will Susi need to buy?

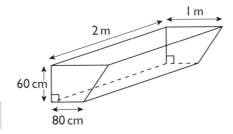

1 m³ = 1000 litres

PB
ES
8 Rhodri has a tank in his garden to store heating oil. The tank is in the shape of a square prism. The tank measures 60 cm by 1.2 m by 1.2 m. The level of oil in the tank is 30 cm from the bottom. Rhodri orders 700 litres of heating oil.

Will there be enough room in the oil tank for all the oil? You must explain your answer.

PB
ES
9 A can of baked beans has a height of 10 cm and a radius of 3.7 cm. The average space needed for one baked bean is 0.2 cm³.

Estimate the number of baked beans in the tin.

PB
ES
10 Zigi has a swimming pool in the shape of a cylinder. The swimming pool has a diameter of 1.2 m. The height of the swimming pool is 90 cm. Zigi empties the swimming pool at a rate of 5 litres of water per minute. She starts emptying the pool at 8.30 a.m.

At what time will the swimming pool be completely empty?

Geometry and Measures Strand 6 Three-dimensional shapes Unit 6 Enlargement in 2 and 3 dimensions

PS **1** Here are two circles. One circle has a radius of 5 cm.
The other circle has a diameter of 5 cm.

    a Find the ratio of the diameters.

    b Work out the ratio of their areas.

PS **2** A block of flats is in the shape of a cuboid. Nathan makes a scale
model of the block of flats. He uses a scale of 1 cm to represent 50 cm.

    a The height of the block of flats is 30 m. What is the height of
the model?

    b The area of the front of the model is 300 cm². What is the area of
the front of the block of flats?

    c The model has dimensions 10 cm by 25 cm by 30 cm.
Find the volume of the block of flats.

PB
ES **3** Mario has a small photograph that he is going to enlarge on his
computer. The length of the small photograph and the length of
the enlargement are in the ratio 2 : 5.

    a Find the perimeter
of the enlargement.

    b Find the area of the
enlargement.

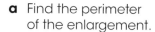

4 cm

← 6 cm →

DF **4** This cuboid has a surface area of 484 cm².

9 cm
8 cm

    a Find the height of the cuboid.

    b Find the volume of the cuboid.

    The cuboid is enlarged by scale factor 5.

    c Find the surface area of this cuboid.

    d Find the volume of this cuboid.

PB
ES
5 The large watering can is an enlargement of the small watering can. The large can is twice the height of the small can. It takes Susan 30 seconds to fill the small watering can.

How long will it take to fill the large watering can using the same tap?

PB
ES
6 Lesley is designing a banner to advertise a concert. He designs this banner on his computer. The full size banner will fit across the doorway of the venue. The doorway has a width of 3.75 m. The banner costs £26 per square metre to make.

Work out the cost of the banner.

Concert
here Today

3 cm

←— 7.5 cm —→

PB
ES
7 Keith is making a model of a football stadium using a scale of 1 : 1000. The area of the base of the stadium in Keith's model is 2500 cm².

    a Find the area of the base of the actual stadium.

    The volume of the actual stadium is 10 000 000 m³.

    b Find the volume of Keith's model.

Geometry Strand 6 Three-dimensional shapes
Unit 7 Constructing plans and elevations

PS — PRACTISING SKILLS   DF — DEVELOPING FLUENCY   PB — PROBLEM SOLVING   ES — EXAM-STYLE

PS **1** Sketch the plan, front elevation and side elevation of these shapes. ●●○

a   b   c   d

PS **2** The diagram shows the plan and elevations of a plinth. ●●○
Make an isometric drawing of this shape.

Plan   Front elevation   Side elevation

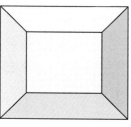   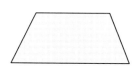   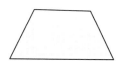

DF **3** The diagram shows a building. ●●○
ES Draw the plan, front elevation and side elevation.

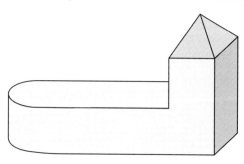

PB
ES

4 The diagram shows Emma's workshop.

The apex of the roof is central to the base.

The maximum height of the workshop is 3.5 m.

Draw an appropriate elevation to scale and use it to work out the area of the roof of the workshop.

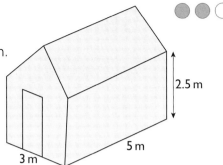

2.5 m

5 m

3 m

DF **5** The diagram shows an isometric scale drawing of a 3D shape.
The isometric grid is made with 1 cm triangles. 1 cm represents 5 cm.

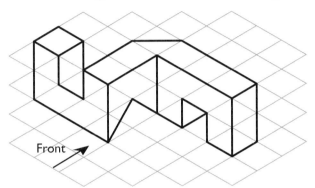

Front

Make an accurate drawing of the plan, front elevation and side elevation of the shape.

PB **6** Here are the plan and elevations of a structure made from toy building bricks.

The structure is made from:

- two cuboids with a square cross-section of side 4 cm and height 2 cm
- one cuboid with a square cross-section of side 2 cm and length 4 cm
- a square-based pyramid with a vertical height of 2 cm.

Draw the 3D shape on isometric paper.

Plan

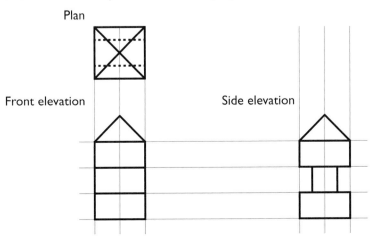

Front elevation

Side elevation

Statistics and probability
Strand 1 Statistical measures
Unit 4 Using grouped frequency tables

PS ─ PRACTISING SKILLS   DF ─ DEVELOPING FLUENCY   PB ─ PROBLEM SOLVING   ES ─ EXAM-STYLE

PS
ES

1 The table shows information about the weights of 50 onions.

Weight (grams), w	Frequency, f	Midpoint, m	$f \times m$
$70 < w \leqslant 90$	12	80	960
$90 < w \leqslant 110$	23		
$110 < w \leqslant 130$	10		
$130 < w \leqslant 150$	5		

     a In which group does the median lie?

     b Copy and complete the table.

     c Find an estimate for the mean weight.

PS
ES

2 The number of complaints received by a TV company on each of 25 days is summarised in the table.

Number of complaints	Frequency, f	Midpoint, m	$f \times m$
0–2	3	1	3
3–5	11		
6–8	7		
9–11	4		

     a Copy and complete the table.

     b Find an estimate for the mean number of complaints.

     c Find an estimate for the range.

DF **3** In an experiment, some children were asked to do a jigsaw puzzle. The table shows information about the times taken to do the jigsaw puzzle.

Time (minutes), t	Frequency, f	Midpoint, m	$f \times m$
$5 < t \leqslant 7$	6		
$7 < t \leqslant 9$	18		
$9 < t \leqslant 11$	13		
$11 < t \leqslant 13$	8		
$13 < t \leqslant 15$	5		

a Write down the modal group.

b How many children did the jigsaw puzzle?

c In which group does the median lie?

d Copy and complete the table.

e Find an estimate for the mean time.

DF **4** In a land survey the areas of 90 fields are measured. The results are summarised in the table.

Area (hectares), a	Frequency, f
$0 < a \leqslant 10$	9
$10 < a \leqslant 20$	15
$20 < a \leqslant 30$	23
$30 < a \leqslant 40$	28
$40 < a \leqslant 50$	15

a Work out the number of fields with an area

  i greater than 30 hectares

  ii 40 hectares or less.

b Write down the modal group.

c Estimate the mean area of the fields.

PB **5** There are 100 trees in Ashdown Woods. The table gives information about the
ES heights of 85 of these trees.

Height (metres), h	Frequency, f
$0 < h \leqslant 4$	30
$4 < h \leqslant 8$	24
$8 < h \leqslant 12$	15
$12 < h \leqslant 16$	12
$16 < h \leqslant 20$	4

Here are the heights, in metres, of the other 15 trees.

3.5	10.3	11.4	6.7	3.9
4.2	12.5	2.4	15.8	17.0
9.5	8.9	14.9	15.2	7.8

a Draw and complete a frequency table for all 100 trees.

b Write down the modal group.

c Find an estimate for the mean height of the 100 trees.

PB **6** The manager of a shoe shop recorded the amounts of money
ES spent on shoes in one day. Her results are summarised in the table.

Amount spent (£), a	Frequency, f
$0 < a \leqslant 25$	4
$25 < a \leqslant 50$	26
$50 < a \leqslant 75$	63
$75 < a \leqslant 100$	17

a Work out an estimate for the total amount of money spent on
shoes that day.

b Find an estimate for the mean.

c Explain why this is only an estimate of the mean.

PB **7** Mrs Abdul takes some children to a theme park. The tables
ES give information about the heights of the children.

Boys	
Height (cm), h	Frequency, f
$120 < h \leqslant 125$	0
$125 < h \leqslant 130$	3
$130 < h \leqslant 135$	7
$135 < h \leqslant 140$	16
$140 < h \leqslant 145$	9

Girls	
Height (cm), h	Frequency, f
$120 < h \leqslant 125$	1
$125 < h \leqslant 130$	3
$130 < h \leqslant 135$	8
$135 < h \leqslant 140$	15
$140 < h \leqslant 145$	8

a Compare the mean heights of the boys and girls.

One of the rides at the theme park has a height restriction.
Children with a height of 130 cm or less cannot go on the ride.

b What percentage of the children cannot go on the ride?

PB **8** The incomplete table below shows some information about the body
ES temperatures of the patients in a hospital.

Temperature (°C), T	Frequency, f	Midpoint, m	$f \times T$
$36.25 < T \leqslant 36.75$	15	36.5	
$36.75 < T \leqslant 37.25$	19		703
$37.25 < T \leqslant 37.75$		37.5	450
	10	38	380
$38.25 < T \leqslant 38.75$	4		

a Copy and complete the table.

b Find an estimate for the mean body temperature of the patients.

c In which group does the median lie?

A patient has a fever if they have a body temperature greater
than 38 °C.

d Find the best estimate for the number of patients with a fever.

PB **9** Balpreet recorded the lifetimes in hours, h, of some batteries. Here are her results.
ES

11.0	10.5	14.2	16.3	18.3	14.8	15.8	12.5	17.9	13.9
15.5	16.3	15.4	17.7	12.3	19.5	13.6	16.8	14.3	14.9
14.5	12.8	13.6	17.6	15.0	14.2	19.6	15.7	13.1	15.7

a Work out the mean lifetime of the batteries.

b Draw and complete a grouped frequency table for this data
using the intervals $10 < h \leqslant 12$, $12 < h \leqslant 14$, etc.

c Use your grouped frequency table to calculate an estimate for
the mean lifetime of the batteries.

d Is your estimate an overestimate or an underestimate of the mean
lifetime of the batteries? Explain why.

Statistics and Probability
Strand 1 Statistical measures
Unit 5 Interquartile range

PS PRACTISING SKILLS   **DF** DEVELOPING FLUENCY   **PB** PROBLEM SOLVING   **ES** EXAM-STYLE

PS **1** For each of these data sets, find
  i the median   **ii** the interquartile range.
  a −8, −5, −3, −2, −2, 0, 3, 5, 7, 8, 10
  b 3.6, 2.7, 4.8, 1.6, 8.3, 7.9, 6.8, 5.4, 3.3, 5.7, 7.0

PS **2** The box plot gives information about the lengths, in mm, of some worms.

Find

  a the median length

  b the interquartile range of the lengths.

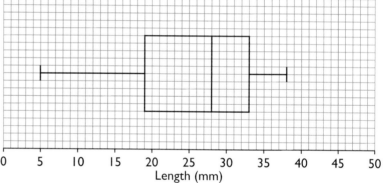

Length (mm)

DF **3** The table gives information about the time taken, in minutes, to serve each of 80 customers at a supermarket check-out.

Time taken (t minutes)	$0 < t \leqslant 2$	$2 < t \leqslant 4$	$4 < t \leqslant 6$	$6 < t \leqslant 8$	$8 < t \leqslant 10$	$10 < t \leqslant 12$
Frequency	7	8	15	23	20	7

  a Draw a cumulative frequency diagram for this information.

  b Find an estimate for
    i the median time   **ii** the interquartile range.

The shortest time taken to serve a customer was 0.5 minutes.
The longest time taken to serve a customer was 11 minutes.

  c Draw a box plot for the distribution of the times taken to serve these customers.

DF **4** The incomplete box plot gives some information about the weights, in kg, of some dogs. The diagram shows the lower quartile, upper quartile and highest weight. The median weight is 8 kg more than the lower quartile.

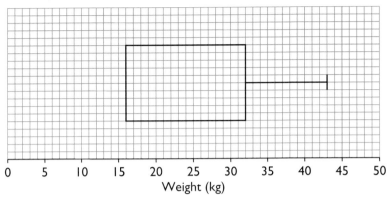

a Work out the median weight.

The lowest weight is 25 kg less than the upper quartile. Work out:

b the lowest weight

c the range of the weights

d the interquartile range.

PB
ES **5** The box plots show information about the average number of miles per gallon (mpg) achieved by a sample of cars in 1990 and in 2010.

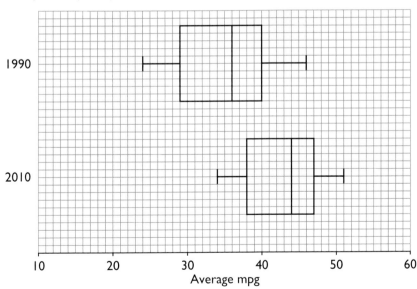

a Write down
  i the highest average mpg in 1990
  ii the lowest average mpg in 2010.

b Compare the medians and the interquartile ranges of the average mpg of these cars.

PB **6** The cumulative frequency diagram gives information about the
ES times taken by some children to do a test.

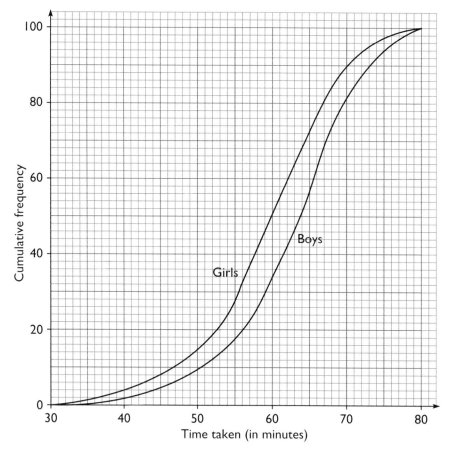

a More girls took 50 minutes to do the test than boys. Estimate how many more.

b Work out the percentage of boys who took more than 70 minutes to do the test.

c Compare the medians and the interquartile ranges of the times taken by these boys and girls to do the test.

Statistics and probability
Strand 2 Statistical diagrams
Unit 3 Pie charts

PS — PRACTISING SKILLS   DF — DEVELOPING FLUENCY   PB — PROBLEM SOLVING   ES — EXAM-STYLE

PS 1 The pie chart shows information about the weights of the ingredients used to make a cake.

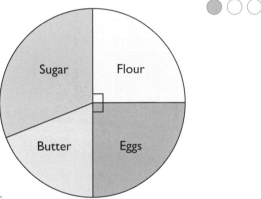

   a Which ingredient has

      i the smallest weight

      ii the largest weight?

   b What fraction of the cake is flour?
      Tony makes this cake. He uses 125 grams of eggs.

   c Work out the total weight of the cake.

PS 2 The pie chart shows information about the finances of a supermarket.

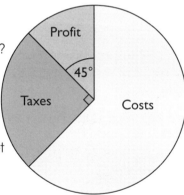

   a What fraction of the finances is profit?
      The angle for taxes is 90°.

   b Work out the angle used for costs.
      The taxes paid were £5000.

   c Work out

      i the profit made by the supermarket

      ii the supermarket's costs.

DF 3 The table gives the numbers of votes that Simon, Helga and Bik got in an election.

	Simon	Helga	Bik
Number of votes	15	6	9

   Tania is going to draw a pie chart to show this information.

   a How many degrees should she use for one vote?

   b Work out the angle she should use for

      i Helga   **ii** Simon.

   c Tania says: 'Bik got more than a quarter of the votes.' Is she right? Give a reason for your answer.

DF **4** The vertical line chart shows information about the counters in a bag.
A pie chart is to be drawn for this information.

    a How many degrees should be used for each counter?

    b Draw the pie chart.

    c What percentage of the counters are white?

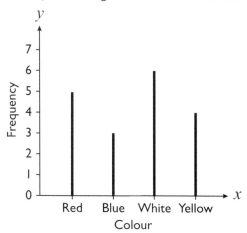

DF **5** Sophie takes photos on her mobile phone. The table shows information
about the numbers and types of photos she took in June.

Type of photo	Family	Holiday	Pets	Friends	Other
Frequency	36	78	60	87	9

Draw a pie chart to show this information.

PB
ES **6** Cleg recorded the numbers and types of trees in a park. His results
are summarised in the pie chart.

    a Measure the angle used for birch.

    b There are 240 trees in the park. Work out the number of birch trees
in the park.

    c There are more sycamore trees than oak trees. How many more?

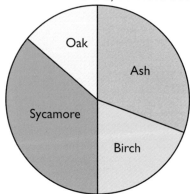

PB
ES
7 In a survey, Phil asked some people what type of music they like the best.
They could choose from pop, jazz and classical. The pie charts give
information about his results.

Phil asked the same number of females and males.

a Use the pie charts to compare the results for the females and
the males. Write down two comparisons.

Phil asked a total of 68 females.

b Work out the number of males who like jazz the best.

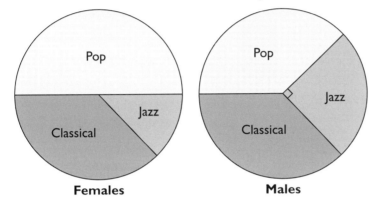

Females   **Males**

PB
ES
8 Sylus recorded the medals won by Great Britain in a competition.
Here are his results.

Gold	Silver	Bronze	Gold	Silver
Bronze	Bronze	Bronze	Bronze	Bronze
Bronze	Silver	Bronze	Silver	Bronze

Draw a pie chart for this information.

PB
ES
9 Some students did a test. The incomplete table and pie chart
give some information about the grades awarded in the test.

Grade	E	D	C	B	A
Frequency			10	16	12
Angle used in pie chart	30				90

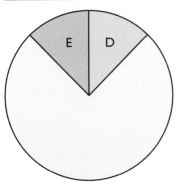

Copy and complete the table and pie chart. Use your table to
complete the pie chart.

Statistics Strand 2 Draw and interpret statistical diagrams
Unit 4 Displaying grouped data

PS PRACTISING SKILLS   **DF** DEVELOPING FLUENCY   **PB** PROBLEM SOLVING   **ES** EXAM-STYLE

PS 1 State whether each type of data is discrete or continuous. The first one has been done for you.

    a The number of wheels on a bus
      discrete

    b The time taken to run 100 metres

    c The mass of an elephant

    d The number of bricks in a wall

    e The temperature of a cup of tea

    f The height of a mountain

    g The number of craters on the moon

PS 2 Copy and complete each category so that it has five equal classes.

    a $0 < w \leqslant 10$ $10 < w \leqslant 20$ _____   _____   _____

    b $100 \leqslant t < 150$ _____ $200 \leqslant t < 250$ _____   _____

    c _____ $15 \leqslant p < 17.5$ _____ $20 \leqslant p < 22.5$ _____

    d $125.7 < d \leqslant 126.2$ $126.2 < d \leqslant 126.7$ _____   _____   _____

    e _____ $2.5 \leqslant c < 2.8$ $2.8 \leqslant c < 3.1$ _____   _____

    f _____   _____ $0.56 < h \leqslant 0.6$ _____ $0.64 < h \leqslant 0.68$

PS 3 Neil has recorded the mass of 30 mice. Here are his results, in grams.

12.7	20.7	15.3	22.8	21.3	18.4	15.9	22.1	19.9	13.5
15.1	19.9	24.7	18.9	14.7	22.0	23.4	18.9	22.4	20.4
20.4	17.2	19.5	17.3	19.1	19.7	17.9	21.8	14.1	16.4

    a Copy and complete the tally chart.

    b Write down the modal class.

Mass, m grams	Tally	Frequency
$12.5 < m \leqslant 15$		
$15 < m \leqslant 17.5$		
$17.5 < m \leqslant 20$		
$20 < m \leqslant 22.5$		
$22.5 < m \leqslant 25$		

4 A doctor recorded the body temperatures of a sample of babies. Some of her results are in this frequency diagram.

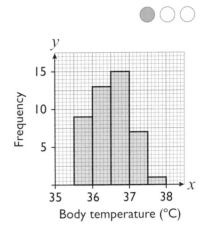

Body temperature (°C)

a Copy and complete the table.

b How many babies were in the sample?

c Work out the percentage of babies in the sample with a body temperature in the range $36 < x \leqslant 37$.

Body temperature, x °C	Frequency
$35 < x \leqslant 35.5$	0
	9
$36 < x \leqslant 36.5$	
$36.5 < x \leqslant 37$	
$37.5 < x \leqslant 38$	1

 5 Monica recorded the time taken, in seconds, to serve individual customers in her shop. Her results are summarised in the frequency table.

Time, t seconds	Frequency
$0 < t \leqslant 10$	4
$10 < t \leqslant 20$	8
$20 < t \leqslant 30$	17
$30 < t \leqslant 40$	12
$40 < t \leqslant 50$	9

a It took more than 40 seconds to serve some customers. How many?

b It took 17.5 s to serve Mr Brown. Which group is he in?

c Draw a frequency diagram to show the data.

 6 The stem-and-leaf diagram gives information about the heights, in metres, of 25 sycamore trees in a wood.

```
0 | 3  5  5  6  6  7  8  9  9  9
1 | 0  2  2  5  4  6  8  9
2 | 2  6  6  7  9
3 | 5  7
```

Key: 3|5 means a tree of height 35 m

a Draw a frequency diagram to show the data.

b Describe the distribution. What, if anything, does this tell you about the ages of these sycamore trees?

PB
ES

7 Amod recorded the lung capacities of 30 adult males. Here are his results, in litres.

5.6	5.7	5.1	5.9	5.5	5.6	5.1	5.6	6.8	5.3
6.1	5.4	6.2	6.4	5.4	5.7	6.4	6.5	5.9	6.4
5.8	5.9	6.3	5.1	5.8	6.6	5.6	5.3	5.8	6.8

   a Make a grouped frequency table for this data, using four equal class intervals.

   b Which class contains the median?

   c Draw a frequency diagram to show the data.

PB
ES

8 Ori recorded the lengths of time, in seconds, some students could stand on their left leg. His results are in this frequency table.

Time (for left leg) t seconds	Frequency
$50 < t \leqslant 100$	12
$100 < t \leqslant 150$	33
$150 < t \leqslant 200$	20
$200 < t \leqslant 250$	8
$250 < t \leqslant 300$	7

He also recorded the lengths of time, in seconds, that these students could stand on their right leg. The results are in this frequency diagram.

Compare the lengths of time that these students could stand on each leg.

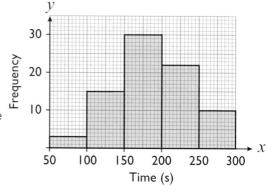

PB
ES

9 Dafydd recorded the masses, m kg, of 300 babies. His results are in this pie chart.

   a Use the information in Dafydd's pie chart to draw a frequency diagram.

   b Cathy says, 'a frequency diagram is a better way to show Dafydd's results'. Do you agree with Cathy? Explain why.

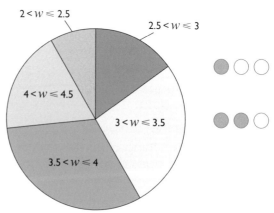

Statistics Strand 2 Draw and interpret statistical diagrams
Unit 5 Scatter diagrams

PS – PRACTISING SKILLS   DF – DEVELOPING FLUENCY   PB – PROBLEM SOLVING   ES – EXAM-STYLE

 1 a Describe the correlation shown in this scatter diagram.

b Draw a scatter diagram to show

i negative correlation

ii no correlation.

 2 The table shows information about the number of pages and the masses of some paperback books.

Number of pages	80	79	80	96	128	72	120	144	95
Mass (g)	135	125	125	129	164	105	155	171	147

a Draw a scatter diagram for this data.

b Describe the correlation in the scatter diagram.

c Draw a line of best fit.

d The paperback *Overlord* has 100 pages. Use your graph to estimate the mass of this paperback.

e *Superfast* has a mass of 160 g. Use your graph to estimate the number of pages in this paperback.

PS **3** The scatter graph shows the prices and mileages of all the used cars at A1 garage.

    a How many used cars are there at A1 garage?

    b What is the mileage of the car with the
        i highest price

        ii lowest price?

    c Work out the mean price of all the used cars at A1 garage.

    d Describe the correlation in the scatter diagram.

    e The garage receives another used car. Its clock shows 65 000 miles. Estimate the selling price of this car.

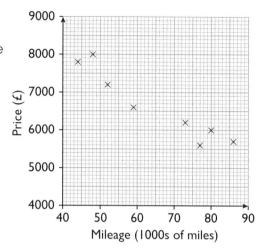

DF
ES
4 Gwilym recorded the heights and the foot lengths of the other seven athletes in his squad. His results are shown in this table.

Height (cm)	148	152	154	158	163	168	160
Foot length (cm)	20.5	23	21	24	25	26.5	25

    a Draw the scatter diagram.

    b Describe the correlation in the scatter diagram.

    c Gwilym's height is 165 cm. Draw a line of best fit and use it to estimate Gwilym's foot length.

DF
ES
5 A scientist measured the density and the speed of sound in eight different gases. The scatter diagram shows her results.

    a For one of these gases, the speed of sound is 349 m/s. What is the density of this gas?

    b Describe the correlation shown in the scatter diagram.

    c Draw a line of best fit and use it to predict the speed of sound in a gas that has a density of 1.2 kg/m³.

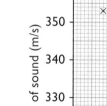
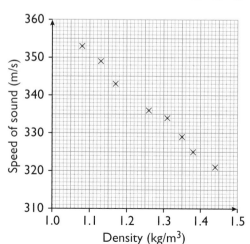

6 State which type of correlation (positive or negative) is suggested by
the following relationships.

   a As the height of an elephant increase, so does its mass.

   b As the age of a tadpole increases, the length of its tail decreases.

   c The greater the amount of fuel in a car's fuel tank, the greater the distance it will travel.

   d The more time people spend working, the less free time they'll have.

   e The greater the height of a weather balloon, the lower the surrounding air temperature.

   f The greater the time spent revising, the greater the number of marks achieved in the test.

 7 The table shows information about the body lengths and the
wingspans of six British birds.

Bird	A	B	C	D	E	F
Body length (cm)	50–60	25–35	40–45	28–40	60–66	63–65
Wingspan (cm)	100–150	60–65	95–115	60–80	145–165	120–150

   a Copy and complete the table below for the mid body lengths and mid wingspans for these birds. The first one has been done for you.

Bird	A	B	C	D	E	F
Mid body length (cm)	55					
Mid wingspan (cm)	125					

   b Draw the scatter diagram.

   c Describe and interpret the correlation between mid body length and mid wingspan.

 8 The scatter graph shows information
about the number of people visiting
an exhibition and the number of
brochures sold in the exhibition
shop each day last week.

   a 45 brochures were sold on
Wednesday.

   How many people visited the
exhibition on Wednesday?

   b 250 people visited the exhibition
on Monday.

   Twice as many people visited the
exhibition on Saturday as on Monday.

   Work out the difference between
the numbers of brochures sold on Saturday and Monday.

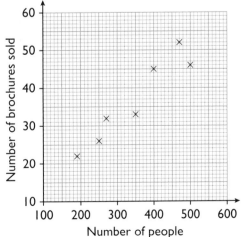

c Describe the correlation in the scatter diagram.

d Steffan says, 'the day on which the median number of people visited the museum is the same day as the day on which the median number of brochures was sold'.

Is he right? Show how you get your answer.

PB
ES

9 The scatter diagrams show information about the marks given to each of eight cakes, A–H, by three judges in a baking competition.

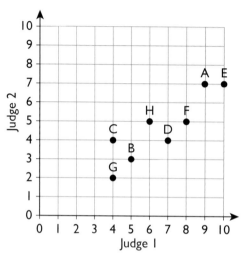
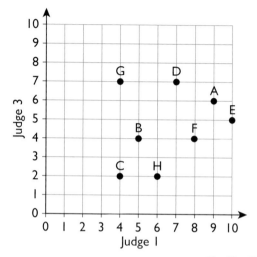

a How consistent are the three judges in their marking? Give reasons for your answer.

b Draw a third scatter diagram to show the marks given to the eight cakes by judges 2 and 3.

c **i** Describe the correlation in this scatter diagram.

ii Did you expect this answer? Explain why.

Statistics Strand 2 Draw and interpret statistical diagrams
Unit 6 Using lines of best fit

PS – PRACTISING SKILLS   DF – DEVELOPING FLUENCY   PB – PROBLEM SOLVING   ES – EXAM-STYLE

PS **1** One of the points in this scatter diagram is an outlier.
Copy the scatter diagram and circle the outlier.

●○○

PS **2** Choose the most appropriate word from the list below to complete
each sentence.

causation   negative   positive   interpolation
extrapolation   outlier   bivariate   trend

●○○

a _____ is when you estimate a value beyond the range of the data.

b A scatter diagram is used to show the relationship in _____ data.

c An _____ is a pair of values that does not fit the overall trend.

d Correlation does not prove _____.

e _____ is when you estimate a value within the range of the data.

DF **3** The table shows the heights and masses of ten players in a
ES football team.

●○○

Player	1	2	3	4	5	6	7	8	9	10
Height (cm)	177	138	184	182	180	178	176	172	170	169
Mass (kg)	114	121	119	117	116	116	115	111	108	110

a Which player is the shortest?

b Draw a scatter diagram for this data.

c Draw a line of best fit.

d Helmut is also in this football team. His height is 175 cm. Use your line of best
fit to estimate the difference between Helmut's mass and the mass of the
heaviest player.

PS **4** This scatter diagram shows the distances and costs of nine train journeys.

A line of best fit has been drawn on the scatter diagram.

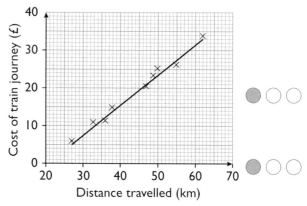

a Tegid travels on a train journey that costs him £17.05. Use the line of best fit to estimate the distance he travelled.

b Sophie is going to travel 58 km by train. Use the line of best fit to estimate the cost of Sophie's journey.

c Jim is going to travel 70 km by train. Use the line of best fit to estimate the cost of Jim's journey.

DF **5** The scatter diagram shows the number of traffic cameras and the number of speeding fines per year in a particular region of the UK over a nine-year period.

A line of best fit has been drawn on the scatter diagram.

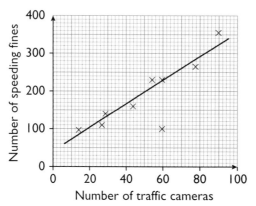

a Identify any possible outliers in the data.

b Describe and interpret the correlation in the scatter diagram.

c Steffan says that traffic cameras cause people to speed. Is he right? Give reasons for your answer.

DF **6** Ulrich measured the wind speed and the air temperature at 1 pm
ES on each of 7 days.
Here are his results.

Wind speed (x km/h)	15.6	8.9	22.1	9.4	0.5	5.8	18.7
Air temperature (y °C)	17.5	20.3	17.8	23.2	19.7	17.5	19.2

a Draw a scatter diagram of Ulrich's measurements.

b Is there any correlation between wind speed and air temperature? Give a reason for your answer.

c A weather forecaster says that the wind speed at 1 pm tomorrow will be 25 km/h.

Ulrich says he is going to draw a line of best fit on the scatter diagram and use it to estimate the air temperature for a wind speed of 25 km/h.

Comment on the reliability of Ulrich's estimate.

PB **7** Charlie is investigating how long it takes paint to dry. He conducts
ES eight experiments in which he paints a wall and records the time it takes for the paint to dry and the average air temperature during this period. Here are his results.

Experiment	Time	Temperature
1	12 hours	3.6 °C
2	6 hours	6.8 °C
3	5 hours 15 minutes	12.9 °C
4	4 hours 30 minutes	18.4 °C
5	4 hours	24.1 °C
6	3 hours 30 minutes	28.7 °C
7	3 hours 15 minutes	32.2 °C
8	2 hours 45 minutes	35.3 °C

a Draw a scatter diagram of Charlie's results.

b Identify the outlier.

c Charlie thinks that there is a relationship between the time it takes for the paint to dry and the average air temperature. Is he right? Explain your answer.

PB **8** The scatter diagram shows information about the time taken for each of eight students to write the same text message first with their left hand and then with their right hand.

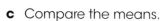

a Mary is one of these students. It took her 17.5 s to write the text message with her right hand. How long did Mary take to write the text message with her left hand?

b Work out the mean time taken for these students to write the text message with their
   i left hand
   ii right hand.

c Compare the means.

d One of these students may be left handed. Which student? Give a reason for your answer.

e Describe the correlation in the scatter diagram.

PB **9** Lorna recorded the resistance of an electronic component at eight different temperatures. The table shows her results.

Temperature (°C)	10	15	20	25	30	35	40	45
Resistance (Ω (Ohms))	3575	3300	2925	2550	2125	1825	1400	1350

a Draw a scatter diagram to show the data.

b Calculate
   i the mean temperature ($\bar{x}$)
   ii the mean resistance ($\bar{y}$).

c Plot the point ($\bar{x}, \bar{y}$) on your scatter diagram.

d Draw a line of best fit on your scatter diagram so that it passes through the point ($\bar{x}, \bar{y}$).

e Use your line of best fit to estimate the resistance of the component at
   i 22.5 °C
   ii 50 °C.

f Which of these two estimates is likely to be more accurate? Explain why.

Statistics and probability
Strand 3 Collecting data Unit 2
Designing questionnaires

PS PRACTISING SKILLS   **DF** DEVELOPING FLUENCY   **PB** PROBLEM SOLVING   **ES** EXAM-STYLE

DF
PB
1 A medical centre is carrying out a survey to encourage patients to
eat five portions of fruit or vegetables. Here is a section from the survey.

Do you eat 5 portions of fruit or vegetables a day?

Yes ☐

No ☐

Occasionally ☐

   a Write a criticism of the question.

   b Write a criticism of the response options.

   c Write a suitable question that could be asked and give appropriate
response boxes.

DF
PB
2 Write a question, with a selection of answer boxes, to find out peoples'
single favourite sandwich filling.

DF
PB
3 Lois is conducting a survey to find out the views of local people about the
council's policy on recycling garden waste. She asks people in the town centre
one Tuesday morning at 12 noon. Here are the first two questions.

1	How much garden waste do you have each week?
2	Do you agree that the council should charge for the collection of garden waste?

Make 3 criticisms of Lois's survey.

PS
DF
4 'People brushing their teeth 3 times a day have fewer fillings.'

   a Do you think this is a reasonable hypothesis? Give a reason for your answer.

   b How would you test this hypothesis?

PS
ES

5 Bryn wanted to find out if people supported the local rugby team.
He carried out a survey outside the local rugby club before a game.
Here is his questionnaire.

1	How old are you?
	16 to 20 ☐
	20 to 30 ☐
	30 to 40 ☐
	Over 40 ☐
2	How often do you come to rugby games?
	Never ☐
	Sometimes ☐
	Often ☐
3	Do you support this rugby club?
	Yes ☐
	No ☐

a Explain why Bryn's survey could be biased.

b Write down a criticism of each question in the questionnaire.

c Rewrite the questionnaire to improve it, addressing all your criticisms.

6 'People are involved in more accidents walking in the street if they
text as they walk through the street than if they don't.'

a Do you think this is a reasonable hypothesis? Give a reason for your answer.

b How would you test this hypothesis? Write any questions you may be asking.

Statistics and probability
Strand 4 Probability Unit 2
Single event probability

PS — PRACTISING SKILLS   **DF** — DEVELOPING FLUENCY   **PB** — PROBLEM SOLVING   **ES** — EXAM-STYLE

PS **1** Here are some cards. Each card has a shape drawn on it. Stephanie ⬤◯◯
is going to take one of the cards at random.

   a Which shape has the greater probability of being taken, an arrow
or a heart? Give a reason for your answer.

   b Write down the probability that the shape will be an arrow.

PS **2** Jasmine rolls a biased die. The probability that the die will land on ⬤◯◯
a 6 is 0.4.

   Work out the probability that the die will not land on a 6.

PS **3** There are 3 red counters, 2 green counters and 6 yellow counters in ⬤◯◯
a bag. Yuan is going to take, at random, a counter from the bag.

   a What is the probability that the counter will be

     i red

     ii green

     iii yellow

     iv white?

   b What is the probability the counter will **not** be

     i red

     ii green

     iii yellow

     iv white?

DF **4** The probability of event A is $\frac{1}{3}$. The probability of event B is 0.35. ⬤◯◯

The probability of event C is 30%.

Write these events in order of likelihood. Start with the least likely event.

169

DF **5** The vertical line chart shows the flavours and numbers of sweets in a box. Maja is going to take, at random, a sweet from the box.

a Which flavour of sweet has the smallest probability of being taken?

b Write down the probability that the flavour will be
  i strawberry
  ii lime.

c Write down the probability that the flavour will not be
  i orange
  ii lemon.

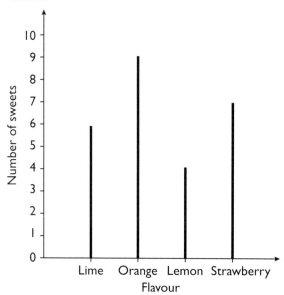

DF **6** The pie chart gives information about the ages of the people watching a film at a cinema.

One of these people is picked at random. Find the probability that the age of this person will be

a 51 years and over

b 50 years or less

c 31–50 years

d 11–30 years.

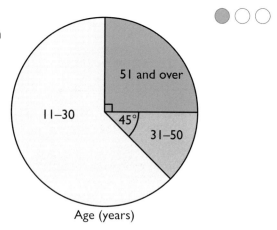

Age (years)

DF **7** The stem and leaf diagram gives the number of people
on each of 25 tours.

1	7 7 8 9 9 9
2	1 2 3 3 3 4 4 6 6 6 8 9 9
3	0 0 1 2 2 3

Key: 1|7 represents 17 people

Henry picks, at random, one of these tours. What is the probability
that this tour has on it

a exactly 26 people

b 19 people or less

c more than 25 people

d between 20 and 25 people?

PB **8** Katie is designing a fair spinner to use in her probability lessons.
She is going to spin the spinner once.

Copy and complete the spinner for each question.
You must only use letters A, B and C.
Write 4 letters on the spinner so that it is

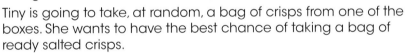

a more likely to get an A than a B

b equally likely to get an A or a C

c twice as likely to get a B than an A.

PB **ES** **9** Box A contains 3 bags of ready salted crisps and 4 bags of cheese
and onion crisps. Box B contains 4 bags of ready salted crisps and
7 bags of cheese and onion crisps.

Tiny is going to take, at random, a bag of crisps from one of the
boxes. She wants to have the best chance of taking a bag of
ready salted crisps.
Which box should she use? Explain why.

PB **ES** **10** Silvia is designing a probability experiment. She puts 15 green
counters and 35 blue counters in a bag.

a What is probability of taking, at random, a green counter from
the bag?

Silvia puts some more blue counters in the bag. The probability
of taking at random a green counter from the bag is now 0.25.

b How many blue counters did she put in the bag?

Statistics and probability
Strand 4 Probability Unit 3
Combined events

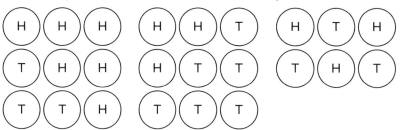

PS — PRACTISING SKILLS   **DF** — DEVELOPING FLUENCY   **PB** — PROBLEM SOLVING   **ES** — EXAM-STYLE

PS **1** Nathalie spins three fair coins. Here are the possible outcomes.

(H)(H)(H)   (H)(H)(T)   (H)(T)(H)

(T)(H)(H)   (H)(T)(T)   (T)(H)(T)

(T)(T)(H)   (T)(T)(T)

Write down the probability that she will get

a 3 heads

b one head and two tails

c two heads and a tail

d at least two heads.

PS **2** Simon goes to a restaurant. He can choose from three types of soup and from three types of bread.

Soup	Bread
Tomato	Brown
Vegetable	White
Lentil	Granary

Simon is going to choose one type of soup and one type of bread.

a List all the possible combinations Simon can choose.

Simon chooses, at random, one type of bread.

b Write down the probability that he will choose Granary.

PS **3** Here are some letters and some numbers on cards.

Safta is going to take, at random, one of the cards with a letter on it and one of the cards with a number on it.

a One possible combination is (A, 2). Write down all the other possible combinations.

b Write down the probability he will take

   i (A, 2)

   ii B and any number

   iii C and a prime number

   iv A or C and an even number.

DF **4** Wilhelm spins a fair 5-sided spinner and a fair 4-sided spinner.

	1	**2**	**3**	**4**
1	2	3	4	5
2	3	4		
3	4			
4				
5				

a Copy and complete the table to show all the possible total scores.

b What is the probability that the total score is

   i exactly 9

   ii exactly 7

   iii an odd number

   iv 4 or less?

DF **5** There are 50 students in a college. Each student may study Polish, Welsh and Chinese.

The Venn diagram gives information about the numbers of students studying all, two, one, or none of these languages.
One of the 50 students is picked at random.

a What is the probability that this student studies

   i all three languages

   ii only Polish

   iii both Chinese and Welsh

   iv Chinese?

173

One of the 17 students studying Welsh is picked at random.

b What is the probability that this student also studies

   i Chinese

   ii both Chinese and Polish?

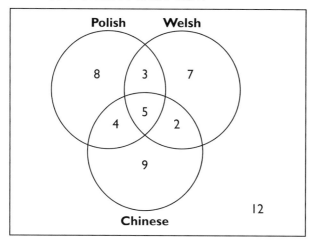

6 Hoeg asked 59 people at a youth club if they play tennis or squash.
35 people said they play tennis. 28 people said they play squash.
17 people said they play both tennis and squash.

a Copy and complete the Venn diagram.

One of these people is picked at random.

b Find the probability that this person plays

   i   squash

   ii tennis, but not squash

   iii tennis or squash, but not both

   iv neither tennis nor squash.

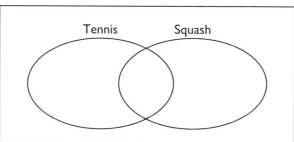

PB **7** Oparbi asked some children which of three films they like the best. The incomplete table shows some information about her results.

	Minions	Inside out	Wall·E	Total
Boys	14	A	8	B
Girls	C	5	D	21
Total	20	E	F	47

a Find the missing values A–F.

One of these children is picked at random.

b What is the probability that this child will be

   i a boy

   ii a girl who likes *Minions* the best.

One of the girls is picked at random.

c What is the probability that this girl likes *Wall·E* the best?

PB **ES** **8** 130 students were each asked to choose an activity for a school trip. The table gives information about these students and the activity they chose.

	Cinema	Theatre	Concert
Male	28	17	14
Female	32	23	16

One of the students is picked at random.

a Find the probability that this student

   i is male

   ii is a female who chose concert

   iii is a male who did not choose cinema.

One of the male students is picked at random.

b Find the probability that this student

   i chose theatre

   ii did not choose concert.

PB **9** Here are some cards. Each card has a letter on it. Helen takes, at random, two of the cards.

a Write down all the possible combinations.

b What is the probability that one of these two cards has S on it?

Jim takes, at random, three of the cards.

c Write down all the possible combinations.

d What is the probability that one of these three cards has S on it?

C A R D S

Statistics Strand 4 Probability
Unit 4 Estimating probability

PS — **PRACTISING SKILLS**   DF — **DEVELOPING FLUENCY**   PB — **PROBLEM SOLVING**   ES — **EXAM-STYLE**

PS 1 Hefin rolls a biased dice 350 times. He rolls 75 sixes. Estimate the probability that Ben rolls a six on his next roll.

PS 2 Bobby throws darts at a dartboard. He is trying to hit the centre of the board. Each time he throws a dart, he records whether he hits the centre (H) or misses the centre (M). Here are his results.

M   M   M   M   H   M   M   M   M   M
M   H   M   H   H   M   M   M   M   M
M   M   M   M   M   M   M   M   M   M
H   M   H   M   H

Estimate the probability that Bobby hits the centre of the dartboard with his next throw.

PS 3 Write down the statistical meaning of each of these terms.

    a population   **b** sample   **c** random sample   **d** trial

DF 4 On her way to work, Heidi passes one set of traffic lights. Over a 30-day period, she had to stop at the lights 16 times.

Heidi thinks she is unlucky with the traffic lights. She says, 'when I drive to work tomorrow, the probability that I will have to stop at the traffic lights is greater than 0.5'.
Is Heidi right? Give a reason for your answer.

DF 5 The table shows the number of games won, lost and drawn for two teams in a handball league.

Team	Games won	Games lost	Games drawn
Stylish Snatchers	12	7	5
Golden Grabbers	18	12	10

    a Which team has played more games?

    b The Stylish Snatchers and the Golden Grabbers are going to play each other in their next game. From the information in the table, which of these two teams is more likely to win the game? Give a reason for your answer.

6 Zoe wants to find the probability that a spider selected at random is female. She looks at five samples of spiders and records the gender of each spider. The table shows her results.

Sample	1	2	3	4	5
Sample size	10	25	50	90	1250
Number of females	8	19	38	68	937
Relative frequency	0.8				

a Copy and complete the table.

b Which of these relative frequencies gives the best estimate for the probability that a spider selected at random will be female? Give a reason for your answer.

c Use the information in the table to find a better estimate for this probability.

7 Alys owns a coffee shop. She offers a free biscuit with each hot drink she sells.
Each customer having a hot drink can choose from a digestive biscuit, a custard cream or a bourbon biscuit.
The two-way table shows some information about the biscuits chosen by 150 customers.

	Digestive biscuit	Custard cream	Bourbon biscuit	Total
Male	25			78
Female		32		
Total	37	48		

a Copy and complete the two-way table.

b Estimate the probability that the next customer that buys a hot drink at the coffee shop will

 i choose a custard cream

 ii be male and choose a digestive.

c Alys says that males are more likely to choose a bourbon biscuit than females.
Is she right? Show how you get your answer.

MATHEMATICS ONLY

PB **8** The table shows the types and numbers of tyres sold by a tyre shop
ES   last week.

Type of tyre	AB303	AC415	XX137	TK700
Number of tyres	36	27	45	58

A customer wants to buy a tyre at the shop.

a Estimate the probability that this customer will buy a TK700 tyre.

b The stock of tyres in the shop is getting low. The manager is going to order a total of 1000 tyres to replace the stock.

How many of each type of tyre should she order?

PB **9** Last week, 109 people donated blood at a clinic. The bar chart shows
ES   information about the number of donors and their blood groups
         (O, A, B and AB).

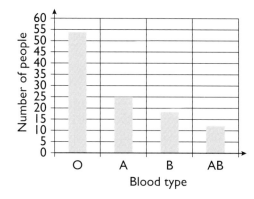

a Estimate the probability that the next person to donate blood at the clinic will have blood type A.

There are a total of 980 people registered to donate blood at the clinic.
The table shows information about the number of donations made by these registered donors during the last year.

Number of donations	0	1	2	3
Number of people	175	525	170	80

b Show that the total number of donations during the last year was 1105.

c Estimate the number of donations of blood type O during the last year. Explain why this is an estimate.

Statistics Strand 4 Probability
Unit 5 The multiplication rule

PS PRACTISING SKILLS   **DF** DEVELOPING FLUENCY   **PB** PROBLEM SOLVING   **ES** EXAM-STYLE

PS **1** A box contains three blue pens and six black pens.

    a Celine takes a pen at random from the box. Write down the probability that the pen will be

      i blue

      ii black.

    She puts the first pen back and takes a blue pen from the box.

    b How many

      i blue pens

      ii black pens

    are now in the box?

    c Celine does not replace the blue pen and takes another pen from the box at random. Write down the probability that the pen is

      i blue

      ii black.

PS **2** Bag A contains 3 red counters and 2 blue counters.

    Bag B contains 2 red counters and 5 blue counters.

    a Copy and complete this tree diagram.

    Catrin takes 1 counter from bag A and 1 counter from bag B without looking.

    b Work out the probability that both counters will be

      i red

      ii blue.

    c Work out the probability that the counter from bag A will be blue and the counter from bag B will be red.

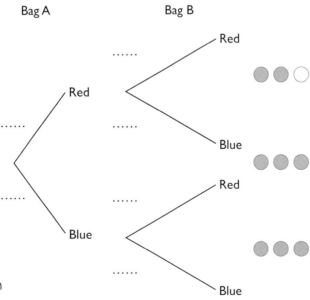

PS **3** Wyn has a fair five-sided spinner like the one in the diagram.
He is going to spin the spinner twice.
Work out the probability the spinner lands on

a A followed by A

b A followed by B

c B followed by C.

Wyn now spins the spinner three times. Work out
the probability the spinner lands on

d A followed by A followed by C.

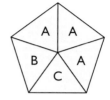

PS **4** A box contains 7 lemon sweets and 6 lime sweets.

Mair takes 2 sweets from the box at random.
Copy and complete the tree diagram.

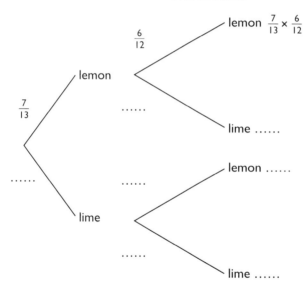

First sweet   Second sweet

lemon $\frac{7}{13} \times \frac{6}{12}$

$\frac{6}{12}$

lemon

$\frac{7}{13}$

......

lime ......

lemon ......

......

......

lime

......

lime ......

DF **5** Delyth has a bag of coins and a box of coins.

Without looking, she takes a coin from the bag and a coin from the box.

The probability that the coin from the bag is a £1 coin is $\frac{4}{7}$.

The probability that the coin from the box is a £1 coin is $\frac{3}{4}$.

a Work out the probability that the coin from the bag and the coin from the box
are both £1 coins.

b Work out the probability that the coin from the bag is a £1 coin and the coin
from the box is not a £1 coin.

c Write down the situation represented by the calculation $\frac{3}{7} \times \frac{1}{4} = \frac{3}{28}$.

6 Tom and Simone both think of a number from 1 to 9 inclusive.

    a Work out the probability that they both think of

      i 3

      ii an even number

      iii a number greater or equal to 7

      iv a prime number.

    b Work out the probability that

      i Tom thinks of a number greater than 3 and Simone thinks of a number less than 5

      ii Tom thinks of a square number and Simone thinks of a prime number.

 7 Zoe takes two tests, A and B.

The probability that she passes test A is 35% and the probability that she passes test B is 85%.

The events are independent.

Work out the probability that Zoe will

    a pass both the tests

    b fail both tests

    c pass only one of the tests.

 8 Dilys and Mica are playing a game. They each need to roll a six on a normal six-sided die to start the game.

    a What is the probability that Dilys will start the game on her

      i first roll

      ii second roll?

    b What is the probability that Mica will start the game on his fifth roll of the dice?

 9 Sri wears a shirt and a tie to work.

The probability that Sri wears a white shirt is 0.8.

When Sri wears a white shirt, the probability that he wears a pink tie is 0.75.

When Sri does not wear a white shirt, the probability that he wears a pink tie is 0.35.

    a Draw a tree diagram to represent this situation. Fill in all the probabilities.

    b Work out the probability that Sri does not wear a pink tie when he goes to work tomorrow.

Statistics Strand 4 Probability
Unit 6 The addition rule and
Venn diagram notation

 **PS** PRACTISING SKILLS  **DF** DEVELOPING FLUENCY 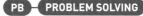 **PB** PROBLEM SOLVING  **ES** EXAM-STYLE

PS **1** The Venn diagram shows information about the numbers of students studying Chinese, Japanese and other languages in a college.

One of these students is chosen at random.

a Work out the probability that this student studies

 i Chinese and Japanese

 ii Chinese or Japanese

 iii only Japanese.

b Work out the probability that this student does not study

 i Japanese

 ii Chinese.

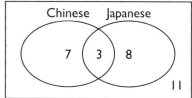

PS **2** **a** Event A and event B are mutually exclusive events.
P(A) = 0.48 and P(B) = 0.37.

 i Draw a Venn diagram showing this information.

 ii Find P(A or B).

b Event C and event D are not mutually exclusive events.
P(C) = 0.8, P(D) = 0.5 and P(C and D) = 0.4.

 i Draw a Venn diagram showing this information.

 ii Find P(C or D).

DF **3** 65% of the people at a party arrived by taxi and 80% arrived in fancy dress.

55% arrived by taxi and in fancy dress.

a Draw a Venn diagram to show this information.

b Work out the probability that a person chosen at random arrived by taxi or wore fancy dress.

4 Giles asks 52 people which, if any, of three coffee shops they go to.
Here are his results.

- 15 people go to coffee shop A.
- 25 people go to coffee shop B.
- 12 people go to coffee shop C.
- 8 people go to coffee shop A and coffee shop B.
- 2 people go into coffee shop B and coffee shop C.
- 0 people go into all coffee shops.
- 0 people go into coffee shop A and coffee shop C.
- 10 people do not go to any of these coffee shops.

a Copy and complete the Venn diagram to show this information.

b Work out the probability that a person chosen at random
   i goes only to coffee shop A
   ii does not go to coffee shop B
   iii goes to coffee shop A or coffee shop B
   iv goes to coffee shop A or coffee shop C.

c Given that a person goes to coffee shop A or coffee shop B, what is the probability that they also go to coffee shop C?

5 There are 120 computers in a showroom.
Each computer has either a 5 GB video card or a RAM extension pack or both of these.
86 have a 5 GB video card.
75 have a RAM extension pack.
Idris picks one of the computers at random.
What is the probability that he picks a computer with both a 5 GB video card and a RAM extension pack?

6 A box contains orange sweets, strawberry sweets, lime sweets and lemon sweets.
The probability of selecting an orange sweet at random is 0.18.
The probability of selecting a strawberry sweet at random is 0.25.
The probability of selecting a lime sweet at random is 0.37.
Work out the probability of selecting, at random

a an orange sweet or a strawberry sweet

b a strawberry sweet or a lemon sweet.

7 $P(A) = 0.3$ $P(B) = 0.8$ $P(A \text{ or } B) = 0.86$
Show that event A and event B are independent events.

DF **8** Make a copy of the Venn diagram to answer each part of this question.

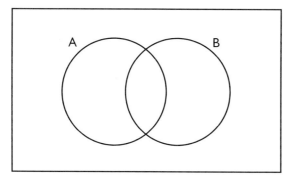

Shade the following regions.

a $A \cap B'$

b $(A \cup B)'$

DF **9** A food scientist is investigating different brands of muesli.

The two main ingredients she is interested in are pumpkins seeds and sesame seeds.
The food scientist analyses 50 different brands of muesli.
She finds 20 of the brands contain neither pumpkin seeds nor sesame seeds.
18 of the brands have pumpkin seeds and 22 of the brands have sesame seeds.

a Show this information on a Venn diagram.

b Estimate the probability that a brand of muesli selected at random will contain both pumpkin seeds and sesame seeds.

c Why is your answer in **b** only an estimate?

PB **10** Make a copy of the Venn diagram to answer each part of this question.

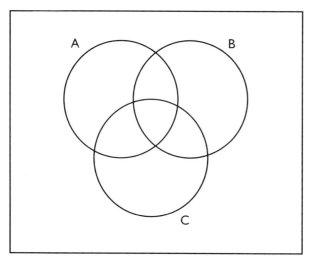

Shade the following regions.

a $(A \cap B) \cup C$ **b** $(A \cup B) \cap C'$